Luban Workshop Series
The 14th Five-Year Plan Series
Textbooks of General Higher Education

普通高等教育"十四五"系列教材

Coastal Engineering
海岸工程

主　编　张　蔚

副主编　季小梅　黄　睿

·北京·

内 容 提 要

本书共分 6 章，内容包括绪论、海岸工程建筑物设计标准、波浪与建筑物的相互作用、海堤工程、防护工程和防波堤工程。

本书适用于港口航道与海岸工程、船舶与海洋工程、海洋资源开发技术、土木工程、环境工程等专业，也可供相关领域涉外工程技术人员参考使用。

图书在版编目（CIP）数据

海岸工程 = Coastal Engineering : 英文 / 张蔚主编. -- 北京 : 中国水利水电出版社, 2024. 8. -- (普通高等教育"十四五"系列教材). -- ISBN 978-7-5226-2614-7

Ⅰ. P753

中国国家版本馆CIP数据核字第20247P4B40号

书　　名	Luban Workshop Series The 14th Five-Year Plan Series Textbooks of General Higher Education 普通高等教育"十四五"系列教材 **Coastal Engineering** **海岸工程**
作　　者	主　编　张　蔚 副主编　季小梅　黄　睿
出版发行	中国水利水电出版社 （北京市海淀区玉渊潭南路1号D座　100038） 网址：www.waterpub.com.cn E-mail：sales@mwr.gov.cn 电话：（010）68545888（营销中心）
经　　售	北京科水图书销售有限公司 电话：（010）68545874、63202643 全国各地新华书店和相关出版物销售网点
排　　版	中国水利水电出版社微机排版中心
印　　刷	清淞永业（天津）印刷有限公司
规　　格	184mm×260mm　16开本　16.25印张　522千字
版　　次	2024年8月第1版　2024年8月第1次印刷
印　　数	0001—1000册
定　　价	**52.00元**

凡购买我社图书，如有缺页、倒页、脱页的，本社营销中心负责调换

版权所有·侵权必究

Synopsis

The "Coastal Engineering" is the core foundational course of the Harbor, Coastal and Offshore Engineering. In order to adapt to the needs of bilingual teaching, based on the domestic curriculum guidelines, we have revised it with reference to the *Coastal Engineering* (Chinese version) edited by Hohai University, to make it suitable for bilingual teaching.

This book comprises six chapters, encompassing an introduction, design standards for coastal engineering structures, the interaction between waves and structures, seawall engineering, protective works, and breakwater engineering, with corresponding exercises appended at the end of each chapter. It has been selected as a key English-language textbook for international students at Hohai University, constituting an integral part of Hohai University's full-English online flagship courses for international students and a offline top undergraduate program of Jiangsu Province. It is suitable for use in bilingual coastal engineering courses offered by higher education institutions in disciplines such as port and navigation engineering, naval architecture and ocean engineering, marine resource exploitation technology, civil engineering, environmental engineering, and other related fields. Additionally, it is also useful for professionals in relevant sectors who engage in foreign-related engineering work.

Preface

"Coastal engineering" refers to various engineering facilities constructed and adopted for coastal protection, exploitation and utilization of coastal zone resources. It mainly includes coastal protection engineering, harbour engineering, estuary management engineering, and marine dredging engineering, which is an important part of ocean engineering. "Coastal engineering" mainly studies the types, functions, features, structural forms, design processes and design methods of coastal engineering such as coastal protection engineering, harbour engineering. Coastal protection engineering mainly protects coastal towns, farmlands, salt farms and shoals to prevent the flooding of storm surges and to resist the invasion and scouring of waves and currents. The main structures include seawall, bank protection and beach protection engineering (including groin dam, groin dam group, offshore embankment, etc.).

"Coastal engineering" is an important course for undergraduates majoring in Harbor, Waterway, Coastal and Ocean Engineering. The most important monograph of coastal engineering in Chinese was written by academician Kai Yen and published in 2002. For the purposes of bilingual teaching, a corresponding textbook of English version is necessary.

This textbook is compiled according to the national syllabus. Chapter 1 gives a general introduction to coastal engineering. Chapter 2 presents the design standards of coastal engineering structures. The emphasis is placed on calculation of the design tidal level and design wave height with different recurrence interval. Chapter 3 is devoted to the analysis of the interaction between waves and structures. Seadike engineering is introduced in Chapter 4 including seadike layout, design of seadike structure and ecological seadike. Chapter 5 presents coastal protection engineering, including revetments, groins, detached breakwaters, artificial beach nourishment and bio-slope-engineering. The cross-section size, structure and optimized design of different types of breakwater engineering are described in Chapter 6.

The material in this book has been used for undergraduate teaching by Shuhua Zhang, Chaofeng Tong, Yanqiu Meng, Jialing Hao, Hongjun Zhao, Yuyang Shao, Yuliang Zhu, Yanwen Xu, Wei Zhang and Xiaomei Ji at Hohai University. Chapter 1 – 3 were written by Wei Zhang, Chapter 4 – 5 were written by Xiaomei Ji, and Chapter 6 was written by Rui Huang. The authors thank the help of Drs. Jun Fan, Ruili Fu, and Yuan Li, and graduate students Hui Wang, Hao Li, Taoning Dong and Sichao Wang from Hohai University.

Wei Zhang
2024. 8

Contents

Synopsis

Preface

Chapter 1　Introduction ……………………………………………………………… 1
　1.1　Coastal zone ……………………………………………………………………… 1
　1.2　Variation influencing factors of coastal erosion and siltation ……………… 14
　Exercise ………………………………………………………………………………… 18

Chapter 2　Design Standards of Coastal Engineering Structures ……………… 20
　2.1　Design standards ………………………………………………………………… 20
　2.2　Design tidal level calculation …………………………………………………… 29
　2.3　Design wave calculation ………………………………………………………… 34
　Exercise ………………………………………………………………………………… 43

Chapter 3　The Interaction between Waves and Buildings ……………………… 45
　3.1　The interaction between waves and sloping buildings ……………………… 46
　3.2　The interaction of waves with a straight wall ………………………………… 80
　Exercise ………………………………………………………………………………… 96

Chapter 4　Seadike Engineering ………………………………………………… 100
　4.1　Management and layout of the dike line …………………………………… 100
　4.2　Types of cross-shore section of seadike ……………………………………… 101
　4.3　Basic size of the seadike section ……………………………………………… 105
　4.4　Construction of seadike ………………………………………………………… 108
　4.5　Design of seadike ……………………………………………………………… 116
　4.6　Ecological seadike ……………………………………………………………… 154
　Exercise ……………………………………………………………………………… 156

Chapter 5　Protection Engineering ……………………………………………… 159
　5.1　Introduction ……………………………………………………………………… 159

5.2 Revetments ··· 163
5.3 Groynes ··· 170
5.4 Detached breakwaters ··· 178
5.5 Artificial beach nourishment ··· 185
5.6 Biological coast protection ··· 190
Exercise ··· 192

Chapter 6 Breakwater Engineering ··· 194
6.1 Introduction ·· 194
6.2 Mound breakwaters ··· 207
6.3 Vertical breakwaters ··· 223
6.4 Mixed breakwaters ·· 236
6.5 Floating breakwaters ·· 238
6.6 Optimal design of breakwaters ······································ 239
Exercise ··· 246

Bibliography ·· 250

Chapter 1　Introduction

1.1　Coastal zone

The coastal zone is an area where the ocean and land meet. It is the most frequent and active place where the four spheres of hydrosphere, lithosphere, atmosphere and biosphere interact with each other in nature. It has the unique environmental characteristics of both ocean and land.

According to the regulations of the national multipurpose investigations of the coastal zone and tidal wetland resources, the range of the coastal zone refers to a narrow strip of land 10 km from the coastline and a shallow sea 15 – 20 m from the sea. The length of coastline and area of coastal zone measured in the national coastal zone survey completed in 1986 are shown in Table 1.1.1.

Table 1.1.1　　The length of coastline and area of coastal zone

Province	Length of continental shoreline /km	Length of island shoreline /km	Islands Number	Coastal zone area/km²							
				Area /km²	Land area	Tidal flat area	Sea area				
							0–5 m	5–10 m	10–15 m	15–20 m	Subtotal
Liaoning	1,971.5	649.0	404	203.4	10,157.6	1,974.2	3,561.8	4,161.8	3,922.2		11,645.8
Hebei	421.0	178.0	107	14.2	3,756.4	1,167.9	1,124.0	1,486.0	1,798.0	2,026.0	6,434.0
Tianjin	153.3	4.2	9	0.2	1,866.2	370.3	847.1	746.2			1,593.3
Shandong	3,122.0	611.4	296	136.6	19,777.0	3,223.6	3,351.9	4,867.5	6,615.4	14,196.5	29,031.3
Jiangsu	953.0	58.4	15	21.7	5,900.6	5,090.4	3,620.4	6,501.2	12,576.3	16,251.5	38,949.4
Shanghai	172.0	277.4	7	1,185.8	3,906.3	904.2	2,341.5	3,112.0	1,061.2	538.9	7,053.6
Zhejiang	1,840.0	4,301.2	1,921	1,670.1	12,100.0	2,444.0	2,272.3	5,326.0	11,055.0		18,653.3
Fujian	3,051.0	1,779.0	1,202	654.0	14,515.0	2,069.0	2,167.0	1,803.0	4,989.0		8,959.0
Guangdong	3,368.1	3,460.8	828	34,804.9	30,469.6	2,530.4	6,442.8	6,320.9	6,563.1	9,229.0	28,555.8
Guangxi	1,083.0	354.5	624	45.8	4,595.7	1,005.3	1,437.6	1,159.0	1,206.4	2,685.3	6,488.3
Total	16,134.9	11,673.9	5,413	38,736.7	107,044.4	20,779.3	27,166.4	35,483.6	94,713.8	157,363.8	
Total coastal zone area /km²	285,187.5										

Chapter 1   Introduction

The coastal zone is generally composed of three parts: supratidal zone, intertidal zone and subtidal zone. In the national coastal zone, the supratidal zone above the mean high tidal level accounts for 38% of the total area, the intertidal zone between the mean high water and low tidal level accounts for 7%, and the subtidal zone below the mean low tidal level accounts for 55%. Fig. 1.1.1 is a typical beach profile diagram, reflecting the coastal geomorphic features of the most intense and significant areas of land and water interaction in the coastal zone.

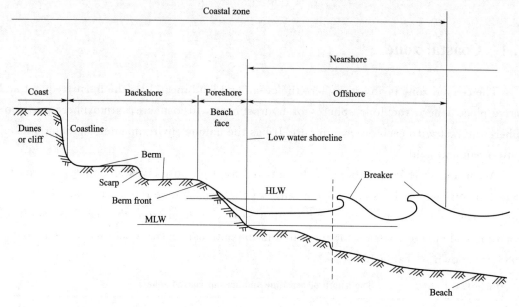

Fig. 1.1.1   Beach profile

1.1.1   Coastal type

National coastal zone is backed by the vast Eurasian continent and faces the endless Pacific Ocean. It is close to four marginal seas and continental shelves of the Bo Hai, the Yellow Sea, the East China Sea and the South China Sea, and has a long coastline. The coastline is generally the boundary where the mean high water annual spring tides meet the land. China's continental coastline stretches from the mouth of the Yalu River in Liaoning Province in the north to the mouth of the Beilun River in Guangxi Zhuang Autonomous Region in the south, totaling more than 18,000 km. The coastline of the island is about 14,000 km. The long coast is "S" shaped lying in the southeast coast of our country, longitudinal across temperate, subtropical and tropical three climatic zones, the length of subtropical coasts accounts for 60% of the total length, most coastline ports are not frozen, four seasons navigation, natural conditions are very superior. Chinese coastal zone covers about 280,000 km² and is an important part of China's territory.

According to its morphology, origin and material composition, Chinese coast can beclassified into five categories: rocky coast, sandy coast, silty coast, mangrove coast and coral reef coast.

1. Rocky coast

The length of China's bedrock coastline is more than 5,000 km, accounting for more than 1/4 of the total coastline, mainly distributed in Liaodong Bandao, eastern and southeastern Shandong Bandao, south of Zhejiang Haijiao, Fujian, Guangdong (the Dapeng Wan, the Daya Wan, the Nanao Dao), west of Fengjiangkou Guangxi and eastern and northern Taiwan, Wanning of Hainan Dao to Sanya and other places. Its main characteristics are as follows: ①The shoreline is tortuous, with alternating promontory and bay, and the erosion and accumulation are interchangeable; ② The bank slope is steep, the bank is very narrow, and the terrain transverse change is significant; ③The coastal dynamic factors are mainly waves, and the sea erosion forms develop at different heights.

Rocky coasts often have deep, well-sheltered waters that make excellent port sites.

2. Sandy coast

The sandy coast is most extensively represented along the shoreline of Liaoning Province, reaching 850 km, accounting for 43% of the province's coastline. They are mainly distributed in parts of the coast of Liaodong Bandao and the west side of Liaodong Wan, the Luanhekou Delta of Hebei Province, the north and south of Shandong Bandao, the north of the Haizhou Wan of Jiangsu Province, Xiuzhen Estuary, the top of some bays of Zhejiang and Fujian, parts of the coast of Guangdong and Guangxi, the west coast of Taiwan, the east and west coasts of Hainan, and the Xisha Qundao. Its main characteristic is that the shoreline is relatively straight and is wider than the bedrock shoreline. The accumulation landform is developed, often accompanied by the form of bar-lagoon composed of shore and offshore dams. The coastal dynamics is still dominated by wave, and the loose material has both lateral movement and longitudinal movement along the coast. This kind of beach is a good place to develop tourism.

3. Silty coast

The silty coast in Jiangsu province occupies the longest coastline, up to 880 km, accounting for 93% of the province's coastline, mainly distributed in Tianjin, the Bohai Wan of Hebei Province, the Yellow River Delta and Laizhou Wan of Shandong Province, the abandoned Yellow River Estuary delta of Jiangsu Province and the radiation bar of Northern Jiangsu, the Yangtze River Estuary of Shanghai, the Hangzhou Wan area of Zhejiang Province, and the north Minjiang River Estuary of Fujian Province. The estuary of the Zhu Jiang in Guangdong and other parts of the coastline, the total length of the coastline is about 4,000 km. According to its landform, the silty coast can be classified into three types: plain type, estuary type and bay type. The plain type is mainly distributed in the Liaodong Wan, the Bohai Wan, the Laizhou Wan and the north Jiangsu coast. The Hangzhou Wan is the most typical estuary type, including the Pearl River

Estuary Lingding Ocean coast. The bay type is distributed in the southeastern coast of the Liaodong Bandao and the Luoyuan Wan in northern Fujian. The main characteristics of the silty coast are as follows: ①The shoreline is straight and the slope is very flat, generally about 1‰; ②The tidal flat is developed with monotonous landform and obvious zonation from land to sea; ③The composition of the substance is fine, with a median grain size of less than 0.06 mm, mainly composed of clay, silty clay, clay sand and sand; ④Tidal current and wave have significant interaction, which is usually dominated by tidal current action, and tidal flat erosion and deposition change frequently. This shoreline beach resources are very rich, which is a good place for aquaculture, land reclamation.

4. Mangrove coast

Mangrove coast is a kind of biocovered coast and mangrove mostly grows in the tropical and subtropical coast between the Tropic of Cancer and the South. The natural boundary of mangrove forests in China is Fuding in Fujian Province. The artificial species can be found in Cangnan (28°N) in Zhejiang Province, and in Keelung, Danshui and Taipei in Taiwan. Mangrove shorelines in Fujian and Guangdong are about 400 km long. The main characteristics of the mangrove coast are as follows: ①Mangrove plants develop best on the silty coast, and can also grow on the depression or mixed deposits of terrigenous debris and biological debris at the back edge of coral reef and on sandy soil; ②Mangrove communities form and have obvious zonation, which can be divided into terrestrial plant zone, semi-mangrove forest zone, beach mangrove forest zone and underwater bank sloping upper zone from land to sea; ③Mangroves have obvious effects of wave elimination, flow retardation and siltation promotion, during which tidal gully system is developed, which is an excellent place for marine creatures to reproduce. The formation of the mangrove coast makes the original coast protected by the flora, so the beach surface is gradually raised and extended, and the new shore area is formed in front of the mangrove, which makes the mangrove coast become a good ecological and silt-promoting sedimentary environment.

5. Coral reef coast

Coral reef coast is mainly distributed along the Nanhai Zhudao, Taiwan and Penghu Liedao and Guangxi. According to the morphology of coral reef, it can be divided into: ①Shore reef, which is distributed along shore; ②Offshore reef, which is separated from land by shallow sea waters; ③Atoll reef, a lagoon surrounded by reefs; ④Table reef, which stands in the ocean and does not form a lagoon; ⑤Uplift reef, the rise of the Earth's crust makes reef surface; ⑥Drowned reef, the growth rate of reef is lower than the crustal settlement rate, so that the reef is addicted to a certain depth. According to the relative distance between coral reefs and the sea surface, it can be divided into: dark beach (dark sand), reef, gray sand bar, gray sand island, reef rock island and so on. The coral reef coast is a habitat for birds and other creatures, and it is also a sea landscape that can develop tourism. Many reef islands are also bases for marine research and defense.

1.1.2 Environmental characteristics of the coastal zone

The coastal zone is the transition zone between the sea and land. The morphological changes of the coastline and the evolution of the landform of the coastal zone are the results of the dynamic action of the land and sea. Therefore, strong Pacific wind waves and tides, changeable continental coastal material composition and structure, huge geographical latitude differences and long-term and frequent human activities are all important factors affecting the natural environment of the coastal zone.

The climate characteristics of coastal zone of China mainly include:

1. Transitional climate under monsoon control

China is a typical monsoon country and coastal climate is mainly subject to monsoon control. Northerly wind prevails in winter, while southerly wind prevails in summer. Spring and autumn are transitional seasons. The Venetian monsoon is mainly formed by the thermal difference between the sea and land. This kind of distribution of cold and heat and the variation of strength and strength determine the process and variation of coastal winter and summer monsoon to a large extent.

The basic characteristics of national coastal climate are:

(1) Monsoon climate features prominently.

In winter, the Asian continent is controlled by cold high pressure from Mongolia, and the northerly wind in front of high pressure becomes the prevailing wind in the east coast of China. Most areas in the coastal zone are cold and dry. The average temperature in January is -9 to -2 ℃ in the Bo Hai coast. From the Yellow Sea to the north of the East China Sea, the temperature is -1 to 4 ℃. Because of low latitude of southern Fujian and Guangzhou-Guangxi coast segment. It is less influenced by winter monsoon. The average temperature of the south coast of the East China Sea in January is 4 to 13 ℃. The average temperature of the south coast is 14 to 15 ℃, but when the north strong cold air moves south, there will be frost damage weather.

In summer, the Asian continent is controlled by low pressure, and the Pacific subtropical high moves north. The east of China forms abundant rainfall under the influence of southeast monsoon. The average temperature of the coastal zone in July is 24 to 29 ℃, and the distribution of temperature isolines is roughly parallel to the coastline, with little difference between the north and the south.

In spring, the weather is changeable, and the rain belt advances gradually from south to north. Southeastern coastal areas are controlled by high pressure in autumn, presenting clear and less cloudy autumn weather.

(2) Transitional climatic characteristics.

Because coastal zone of China is in the transitional zone between Eurasia and the Pacific Ocean, the coastal zone climate has continental and oceanic transitional or mixed characteristics. Under the influence of monsoon, the characteristics of continental and oceanic climate show seasonal variability. In winter, under the influence of strong cold air, the influence of marine climate is weakened, making most of the coast colder. In summer, the southeast monsoon from the Pacific Ocean prevails, and the characteristics of maritime climate are very obvious from south to north, forming the characteristics of rain and heat in the same season.

(3) Meteorological elements change dramatically.

As the coastal zone is a transitional zone of sudden changes of two distinct underlying surfaces of sea and land, there are inevitably characteristics of sharp changes in meteorological elements. The changes of air temperature, water temperature, air pressure and precipitation in the continental or ocean are all distributed in the zonal direction. But they undergo significant changes in the coastal zone, exhibiting distributions parallel to the coastline, and the horizontal gradients of each element increase, forming a zone of sharp variation. Under the conditions of different coast segments, local microclimate may also appear.

(4) Frequent disastrous weather.

Under the influence of monsoon, the spatial and temporal distribution of various meteorological elements such as wind speed, rainfall and pressure is very uneven, and the coastal zone becomes the rapid change zone of these elements, so the coastal area is prone to drought and flood, wind, tornado and other natural disasters. Most of the coastal zone is in the mid-latitude, where the south and north air currents meet, making it more prone to catastrophic weather. Disastrous weather often occurs in the coastal zone, such as cold wave in winter, low temperature rains in spring, typhoons, heavy rains in summer and autumn, and fog and hail in some parts of the coast.

Cold air process temperature drop ≥ 10 ℃ is called cold wave. The number of large-scale cold waves account for 48% of the total number of times in China, and the number of cold waves affecting the south of the Huaihe River accounted for 28% of the total number. March and October to November are the most serious periods of cold waves and strong cold air.

A typhoon is an intense cyclonic storm that occurs over the tropical ocean. The international standard for names and classifications of tropical cyclones are shown in Table 1.1.2.

Table 1.1.2 International standard for names and classifications of tropical cyclones

Name	Tropical depression	Tropical storm	Severe tropical storm	Typhoon
Wind force/level	<8	8 – 9	10 – 11	≥ 12

According to statistics, 210 tropical storms and typhoons made landfall in China during the 30 years from 1951 to 1980, with an average of 7 landfalls per year. More than 90% of them landed in the south of Zhejiang, Taiwan and Hainan Dao, while the number of tropical storms and typhoons landed in the Guangdong Coast section and affected by the eastern coast section accounted for more than 70%. The total daily precipitation is more than 50 mm, which is called heavy rain. The annual rainstorm days are about 2 days in the Bo Hai Coast and 1.5 days in the Liaohe Estuary and the Yellow River Delta, which are the least rainstorm in the coastal zone. 3 to 4 days in the Yellow Sea section, 3 to 5 days in the East China Sea section south of the Hangzhou Wan. The south coast section is mostly in 7 to 10 days. Dongxing of Guangxi has 15.3 days, which is the highest rainstorm in the coastal zone. The rainstorm intensity is the highest in Hainan and Guangxi coastlines, and the lowest in Tianjin and Hebei coastlines. The maximum daily rainfall in the Ledong County, Hainan Province was 962 mm in 1983. The maximum daily rainfalls in Guangxi, Guangdong and Liaoning are above 600 mm, Shanghai and Jiangsu and Zhejiang are above 500 mm, and Tianjin and Hebei are only around 300 mm.

Maximum wind speed $\geqslant$ level 8 (17 m/s) wind call gale, mainly cold wave gale, typhoon gale, thunderstorm gale and so on. The distribution of the annual gale days is more in islands and offshore areas and decreases rapidly from sea to land. The annual number of gale days in the Yellow Sea and the Bo Hai is more than 100 days. Due to the Taiwan Strait narrow tube effect, the east coast section increased to 100 to 180 days, and Lower Dachen Dao topped the coastal zone with 189 days. The annual number of gale days is less the south coast section, which is 40 to 90 days in the islands and only about 10 days in the mainland coast.

2. Zonal and non-zonal distribution of soil and vegetation

Coastal zone of China is located in the middle and low latitudes, spanning three climatic zones from north to south, of which the length of the subtropical accounts for more than half. In the subtropical coastal areas of our country, due to the dominance of the monsoon climate, rainfall is abundant, and for most sections of the coast, the rainy season coincides with the warmest period. The temperature increases from the north to the south, while the annual temperature range decreases from north to south. The temperature difference between the north and the south and the sea water temperature between the north and the south is great in winter and little in summer.

It is precisely due to this combination of heat and moisture in the coastal zone, as well as their harmonious relationship, the vegetation and soil types in the coastal zone are successive: warm temperate deciduous leaf forest-brown soil, brown soil; subtropical evergreen broad-leaved forest-brown soil, yellow red soil and red soil; tropical monsoon rainforest-lateritic soil, lateritic soil. Because there are not many coastal mountains, the differences of vegetation along the vertical elevation are not obvious in the coastal zone. Only the species composition and growth conditions are different due to the differences in hydrothermal conditions. The characteristics of the vegetation are still of the same type as its base level.

There are meadow, marsh and aquatic plants in different latitudes, and the intertidal organisms in different latitudes have the same species in the same tidal area. This indicates that the relative consistency of ecological conditions makes the distribution of plant communities and biological species tend to be trans-zonal and non-zonal, which mainly depends on the soil and microgeomorphic changes of the ecological environment. The complex transitional coastal zone climate makes zonal and non-zonal characteristics coexist, which becomes another characteristic of coastal zone environment.

3. The continents interact strongly with the ocean

The vast territory of China's mainland, with its west high-east low terrain and abundant rainfall, has developed nearly a thousand rivers flowing into the sea, forming major river systems such as the Yangtze River, Yellow River, and Zhu Jiang. The annual average runoff volume reaches 1.6 trillion cubic meters, carrying nearly 17.5 billion t of sediment and 6 billion t of dissolved substances into the sea, providing the coastal zone with a massive amount of terrestrial materials. At the same time, the ocean dynamics in the Pacific are strong. The introduction of Pacific tide is the main power source of China's coastal tide. The introduction of Kuroshio tide is the main power source of national offshore sea current system. The introduction of Pacific wind wave and surging wave is the important power source of our coastal wave. In addition, the introduction of oceanic tsunami is an important cause of abnormal coastal storm surges.

The two different material systems, the continent and the ocean, interact and restrict each other, carrying on the everlasting material transfer and exchange. This complex and long-term process is also significantly influenced by climatic factors, the topography of the sea and the rivers that flow into the sea. Cold waves and monsoons, especially typhoons, have significant effects on coastal waves, currents and storm surges. Sea topography affect tidal wave propagation, forming different tidal wave types and distributions, as well as the distribution of traceback flow field. In particular, the nearshore topography causes wave refraction, diffraction, deformation and fragmentation, which affects the nearshore current system and sediment distribution and transport. A large amount of fresh water and sediment are injected into the sea, which changes the distribution of water temperature, salinity and sediment content in coastal areas. The estuary area of great rivers becomes a special water environment with salt-fresh water mixing, tidal current and runoff interaction. As a result of land-sea interaction, the coastline morphology changes, which also promotes the evolution and development of coastal landform, erosion and accumulation, destruction and shaping, and forms river estuary deltas and alluvial plains and coastal fishing grounds.

The Yellow River carries 1.1 billion t of sediment into the sea every year, two-thirds of which is deposited in its estuarine delta. Due to the huge amount of sediment transport and high concentration of sediment, the estuary of the Yellow River is constantly swinging and wandering, shaping the Bohai Wan and the Laizhou Wan in the Bo Hai. In history, from 1128 to 1855, the Yellow River took the Huai He and poured into the Yellow Sea, forming a Yellow River Delta extending about 100 km in the north of Jiangsu. In 1855, after the Yellow River broke into the Bo Hai again, the abandoned Yellow River Delta in the north of Jiangsu was seriously eroded due to the lack of sediment supply, so far the coastline has retreated more than 20 km. Meanwhile, in the past 100 years, the Yellow River Estuary has been silted into a delta of 2,300 km^2, creating an average of more than 20 km^2 of land every year.

The annual flow of the Yangtze River is 29,000 m^3/s, and about 500 million t of sediment is deposited into the sea every year. Since the end of the Qing Dynasty, the annual new silt area of the Yangtze River Estuary has reached an average of 9 km^2. Since the Yangtze River entered the sea at Zhenjiang and Yangzhou five thousand years ago, the estuary has extended over 100 km, shaping the Yangtze River Delta with an area of 40,000 km^2.

In the rocky coast composed of hills, mountains and platforms of the Liaodong Bandao, the Shandong Bandao and southeastern coastal provinces, under the condition of few terrigenous materials and strong oceanic dynamic action, oceanic erosion cliffs and gravel coast are mostly developed. Some coast sections of eroded materials are moved by coastal currents to form oceanic landforms such as sand dikes, sand spars, lagoons and land islands.

The radial sandbar along the coast of Jiangsu Province is a very unique landform in the muddy coast. The sandbar ranges from the north bank of the Yangtze River Estuary in the south to the Sheyang Estuary in the north, 200 km long from north to south, 90 km wide from east to west. In nearly 2 km^2 of sea areas, more than 70 sand ridges form the sand ridge group with the vertices at the Port of Jiang radiating outward stretch. There are 8 out of water sand ridges and the rest are underwater sand ridges, which are also tidal channels. Radiative sandbar is the material foundation of the underwater delta of the ancient Yangtze River, which is shaped from north to south outside the Qiang Port by the advancing tide wave traveling south to north across the Pacific Ocean and the rotating tide wave reflected off the Shandong Bandao in the Yellow Sea. The radiative sandbar is mainly formed by tidal current and can be called tidal current landform. North and south spring tide system meet in Qiang Port, causing the phenomenon of the convergence and divergence of flood with Hu Port as the center and the tidal surge. The increase of the tidal range and the strength of the tidal current create a good condition for the formation of the offshore sand ridges.

4. Effect of human activity

The environment of coastal zone has been deeply affected by human activities for a long time. The Yellow River Basin, which is the oldest developed area in our country, due to long-term extensive cultivation and thin harvest, reclamation of steep sloping, resulting in serious soil erosion. The Yellow River sediment discharge increased continuously, accelerating the evolution of the estuary and the siltation of the delta. In the Past half century, In the past half century, the ebb and flow of afforestation and soil and water conservation efforts in the upper reaches of the Yellow River have had a direct impact on the rate of siltation and growth of the Yellow River Delta. Due to the construction of reservoirs and sluices in the upstream of the Luanhe River, the sediment into the sea has been greatly reduced, resulting in coarsening and erosion of Tidal Flat at the River Mouth.

Economic construction and resource exploitation in coastal zones, such as port construction, estuarine management, beach reclamation, oil and gas exploitation, construction material mining, sewage discharge and tidal power generation, have a great impact on the coastal environment. In order to protect against coastal erosion, more than 1,000 km of shore protection, seawalls and sea ponds have been built along our coastline, such as the historic Fan Zhongyan Dike, which has made a significant contribution to the coastal prevention. On the contrary, if the project layout is not appropriate, it will cause erosion.

The low water line on the south side of a jetty in Shandong Province receded 100 m within four years after the completion of the jetty, and the high water beach eroded away. This is a profound lesson.

If human activities can be based on science, through investigation and research, unified planning, comprehensive development and utilization, some adverse effects can be controlled and eliminated, and we can obtain the best comprehensive benefits of society, economy and environment. Reeds and rice-grass are planted on the tidal flats in the north and east coast of China, and mangrove plants are protected in the south coast. Good results are achieved in protecting the beach and dikes, reducing waves and promoting siltation and improving the ecological environment.

1.1.3 National coastal resources

The coastal zone is the link between inland and offshore, and it is a unique natural complex, which contains extremely rich natural resources. These resources can generally be divided into three main categories, namely, spatial, material and environmental resources. Spatial resources usually refer to land resources in land and intertidal zone and port resources in water space. Material resources include sea water, fresh water, aquatic products, biological, forest, ore resources, etc. Environmental resources refer to tourism resources and natural reserve resources. As a complex area with concentrated natural resources, the coastal zone is an extremely valuable wealth for the development of national economy.

1. Spatial resources

The coastal zone of our country covers an area of 2.8 million km^2, spanning 18 cities

1.1 Coastal zone

and counties across 11 provinces, municipalities, and autonomous regions along the coast, excluding Taiwan Province. In this stretch of winding coastline, there are nearly a thousand size rivers into the sea, of which 122 rivers are more than 50 km length, the annual inflow of sediment is about 1.75 billion tons. A large amount of sediment is deposited in the estuarine and coastal zone, and the land is silted up to 26,700 – 33,000 hm^2 every year. China's Huanghuai plain in the east, the Yangtze Estuary downstream plain, the Pearl River Delta plain and the lower Liaohe plain, about more than 13.3 million hm^2 of land has historically been raised by sediment silting brought down by rivers. There are now 2.17 million hm^2 of tidal flat in the country and they continue to silt. Growing land resources are a major advantage of the coastal zone.

The water space of national coastal zone also provides excellent conditions for port development. Many large and medium river estuaries, with their deep water and good cover, serve as the gateway for port construction and navigation of Hai He. Many continents and islands have long and winding stretches of coastline, wide water, forming a natural excellent port site. There are more than 150 bays with a coastal area of more than 10 km^2, and the deep-water shoreline of more than 400 km is ice-free. It is estimated that more than 160 ports are available for intermediate berths and above, of which 40 ports are available for berths of 10,000 t or above, and more than 10 ports are available for berths of 100,000 t or above. There is also plenty of room for artificial islands, maritime airports and other offshore facilities.

2. Material resources

The material resources contained in the coastal zone include:

(1) Fresh water resources. China's total freshwater resources have averaged 2.7 trillion m^3 over the years, ranking fifth in the world, but the per capita water consumption is only a quarter of the world's per capita water consumption, and the contradiction between supply and demand of fresh water has become increasingly prominent. The annual average freshwater resources of 11 coastal provinces, municipalities and autonomous regions totaled 746.8 billion m^3, accounting for 27.5% of the country. If the coastal zone is 10 km wide, fresh water resources are about 135.5 billion m^3, accounting for only 5% of the country.

The distribution of fresh water resources along the coast is very uneven. If the water yield per unit area is called the water yield modulus, the water yield modulus of Liaoning, Hebei, Tianjin and Shandong among the 11 coastal provinces, municipalities and autonomous regions is lower than the national average of 285,000 m/km^2; Jiangsu and Shanghai are slightly higher than the national average; Zhejiang, Fujian, Guangxi, Guangdong and Hainan are far higher than the national average. Among them, Guangdong and Fujian reached 3.5 times more than the national average. The per capita water volume of the four northern provinces and municipalities is only 556 m^3, one-fifth of the national per capita water volume of 2,637 m^3, while the per capita water volume of the five southern provinces and regions is 3,840 m^3, about 1.5 times of the national average, but only 1/3 of the world's per capita water volume. Moreover, the total amount of fresh water resources varies greatly from year to year. Compared with the normal year, the fluctuation can reach 15%–20%, and the rainfall in flood season accounts for 70%–80% of the whole year.

(2) Seawater resources. With large reserves of seawater resources, constant composition and significant differences in the concentration of various elements, it is convenient to extract some salts. Sodium chloride is the main component of salt in seawater, accounting for about 80% of all salts, followed by magnesium, sulfur, calcium, potassium, bromine and so on. The salinity is generally about 3.3% (by weight) in national coastal areas except for the estuary area where a large amount of run–off enters the sea with low salinity. At present, the total area of salt pans in China is 360,000 hm^2, and the output of sea salt is 13 million t. By the year 2000, salt chemical products were developed to more than 60 varieties, with a total tonnage of more than 2 million t. The research of seawater desalination technology has also made great progress and entered the stage of production and application.

(3) Marine biological resources. The inshore waters are fertile because they carry a large amount of organic matter with them. In addition, various ecological environments are formed due to the different coastal morphology, sediment and salinity, which provide good conditions for the development of aquatic and aquaculture resources. The area of shallow coastal water city (0–20 m) in mainland China is 162,000 km^2, excluding Taiwan Province and the Nanhai Zhudao. The total amount of fish resources is about 693,000 t with 481 species. Annual catches of coastal fish range from 4 million to 4.7 million t. China's tidal flat area is 2.17 million hectares, about 22,000 km^2. The intertidal zone is a highly productive area of the coastal zone, with an average total biomass of 250 t/km^2, totaling more than 5.2 million t with more than 1,500 species. In 1985, China's mariculture area reached 280,000 hm^2, producing 210,000 t. The coastal ecosystem is very fragile. Over the years, more than 60,000 hm^2 of traditional aquaculture beaches have been reclaimed, leaving a large number of marine life without a living environment. So beach development must be done with caution.

(4) Plant and forest resources. National coastal zone is rich in plant resources, totaling more than 4,000 kinds, among which there are about 1,500 kinds of cash crops, which can be divided into eight categories: medicinal, oil, fiber, starch, timber, forage, tannin and spices. Among them, rice grass is worth mentioning, which has a unique role in coastal protection. Since its introduction from the UK in 1963, rice grass has been cultivated in 33,000 hm^2 after years of promotion. It is salt tolerant, submergence tolerant and multiplies quickly. It plays an obvious role in promoting siltation, building land and protecting beaches. According to the determination of Jiangsu and Zhejiang provinces, the annual siltation capacity of rice grass after planting can reach 6–44 cm. China's coastal forestry land area is 2.64 million hm^2, and the forest coverage rate is 16.6%. Shelterbelts account for only 23% of existing forest species, while timber stands account for

1.1 Coastal zone

43%. In view of the importance of coastal shelterbelt, the area of shelterbelt should be increased to more than 40% in the future. Coastal zone protection forests include seawall protection forest, river dike protection forest, road protection forest, farmland protection forest, water source conservation forest and water and soil conservation forest, sand-fixing forest, and surf forest (mangrove).

(5) Ore resources. Ore resources in the coastal zone, especially oil and gas resources, have good prospects for development. The oil reserves under the sea are estimated at 9 – 13 billion t, and the proven geological reserves of petroleum in the Bohai Wan coastal area alone amount to 3.6 billion t, and natural gas totals 49 billion m^3. In addition, there are 86 kinds of minerals with more than 800 kinds of ore deposits in national coastal zone. They are mainly gold deposits in the Shandong Province, with reserves of about 280 t; Guangdong area ilmenite reserves of 22 million t. The magnesite ore in Shandong has reserves of more than 150 million t. Diamonds are mainly distributed in the Liaoning Province, with an estimated 2,400 kg; Zhejiang-Fujian area of alum stone, reserves of more than 200 million t, ranked first in the world.

(6) Marine energy resources. Marine energy mainly refers to tidal energy, wave energy, temperature difference energy and salt difference energy. China's theoretical reserves on marine energy are 630 million kW, of which 110 million kW tidal energy, annual power generation can reach 275 billion kW · h, mainly distributed in Fujian and Zhejiang provinces, accounting for 81% of the country. The total theoretical power of wave energy is 23 million kW, with more in Shandong, Zhejiang, Fujian and Guangdong provinces, and the theoretical power is between 4 million and 6 million kW. In addition, the vast coastal areas belong to the wind energy rich areas, wind energy resources development potential is quite considerable.

3. Environmental resources

(1) Tourism resources. Coastal tourism resources generally refer to the natural landscape and cultural landscape of the coastline and its adjacent land and ocean (including islands) on both sides. Because national coastal zone spans three climate zones, it has the characteristics of each different climate, reflecting the four characteristics of the beach, seawater, seascape and sea specialty. China's coastal zone is rich in tourism resources. There are more than 1,500 scenic spots that have been developed or to be developed, and more than 100 beaches alone. In the future, it will be developed into six coastal tourism areas, namely the Binhai international tourism zone with Guangzhou as the center, the commercial and cultural tourism zone with Shanghai as the center, the scenic spots and historic sites tourism zone with Hangzhou and Fuzhou as the center, the summer resort along the Bo Hai and the cold resort of Hainan Dao.

(2) Natural reserves. Natural reserves are special areas established by human beings to protect and utilize natural resources and natural environment. As a system, Natural

reserves can be divided into Natural reserves and national parks. The former includes scientific research reserves, natural monuments and natural resources reserves, scenic reserves, etc. The latter includes national parks, forest parks, tourist destinations and so on. China has established 32 Natural reserves for coastal zones and islands, covering an area of about 780,000 hm^2, accounting for about 2.6% of the total area of coastal zones and islands. China's natural reserves can be roughly divided into five categories, namely: natural expansion areas for protecting the forest, vegetation ecological environment mainly, such as Yuntai Mountain, Jiangsu, Liaoning Xingcheng Forest Park; protected areas mainly for the protection of rare animals and plants, such as Jiangsu Dafeng Milu Deer Reserve, Hainan Datianpo Deer Reserve; protected areas designed to protect the coast from erosion, such as the mangrove natural protection bank section of Qinglan Port, Hainan; protected areas for the protecting natural landscapes and historical sites, such as Tianjin ancient coastal relics Shell Embankment Protection area; protected areas for the overwintering of migratory birds and the breeding of aquatic precious animals, such as the domestication of red-crowned cranes in the rare birds Reserve of coastal beaches in Yancheng city Jiangsu province.

1.2　Variation influencing factors of coastal erosion and siltation

The factors that affect the change of coastal erosion and siltation can be divided into two categories: long-term action and short-term action. Coastal changes due to shoreline erosion caused by sea level rise or land subsidence and drastic changes in sediment supply conditions along the coast caused by river diversion belong to long-term factors, which have a large range of influence on the coast and a long period. Short-term action mainly refer to the effects of natural factors such as waves, coastal currents, storm surges, changes in river abundance and variability, wind, and human engineering activities on the coast.

1.2.1　Sea level eustacy

Studies on paleogeology and paleomagnetism show that the short-term reverse polar drift of the earth's magnetic field is often the precursor of the rapid warming of the global climate, resulting in the retreat of mountain glaciers at the poles and middle and high latitudes, and thus causing the rise of the world ocean surface. The Holocene sea flooding is the biggest geological event in the last ten thousand years. It changes the sedimentary pattern of China's shelf sea and controls the changes of China's coastline. The Holocene sea immersion reached its climax 6,000 to 5,000 years ago, about 5 m above present sea level. Since then, the sea level has changed frequently, and shown a trend of fluctuation and rise in the past thousand years. Since the beginning of this century, the earth's average temperature has increased by 1 ℃ compared with the end of the last century, and the

1.2 Variation influencing factors of coastal erosion and siltation

global greenhouse effect has accelerated the rate of sea level rise around the world. In the last century, sea level rose by 1 to 1.5 mm/a, but in the last 50 years, the rate of increase has accelerated to an average of more than 2 mm/a. Data show that the average rate of sea level rise in this century is 0.35 to 0.41 cm/a. In 1985 the National Academy of Sciences estimated that sea levels around the world were likely to rise by 50 – 200 cm by 2100 AD. This change in sea level causes some coastal lands to be submerged year by year.

The most direct impact of sea level rise is the inundation of coastal low-built marshes, which are generally located below the highest annual tidal level and above the mean sea level. For example, nearly half of the land of the Netherlands is located less than 1 m above sea level. For several thousands of years coastal marshes have generally risen in step with the sea level, and as the sea level slowly rises, sediment deposits form new marshes. But if sea level rises faster, the area of marshes will get smaller and smaller.

Rising sea levels will also wash away land above sea level. Because land above sea level is generally steeper than beach terrain, the amount of scour caused by sea level rise is much greater than the amount of land directly submerged. On the West coast of the United States, every centimeter of sea level rise pushes back the coastline by one to four meters.

In addition, rising sea levels increase the base of storm surges and make it difficult to drain lowlands, thus increasing the risk of coastal flooding and storm surge damage. Sea level rise also increases the ability of saltwater to travel upstream in rivers, causing salinization of coastal land and possibly raising the beds of sandy rivers, threatening the safety of shipping.

1.2.2 River migration

The river migration causes the abrupt change of dynamic sediment condition, which affects the balance of erosion and deposition. As mentioned above, in 1128 AD, the Yellow River took over the Huai He and flowed into the Yellow Sea, bringing a large amount of sediment to the northern section of the coast of northern Jiangsu, which rapidly moved the coast eastward. The silt growth rate at the mouth of the Yellow River reached 215 m/a, and the Yellow River Delta of northern Jiangsu was formed in more than 700 years. In 1855, when the Yellow River returned to its northern corse, the northern coast of Jiangsu cut off a huge amount of sediment sources. As a result, the shoreline at the mouth of the abandoned Yellow River retreated at the rate of 1 km/a, and then the rate of erosion slowed down, but it was still 300 – 400 m/a. It was not until the construction of coastal protection projects in recent decades that the shoreline was basically stable. The shoreline has retreated more than 20 km in more than 100 years.

1.2.3 Wave

The changes of natural beaches under the action of waves can be divided into two categories, one related to the movement of sediment perpendicular to the coast or laterally, and the other related to the movement of sediment along the coast or longitudinally.

The change of bank profile caused by the lateral movement of sediment generally shows that sediment is transported back and forth between the offshore area and the shore, so the shoreline only swings back and forth along the average shoreline without any permanent change. In winter, the prevailing storm period, the upper part of the beach is eroded by the wind and waves, and the sediment is transported to the offshore area to accumulate and form the bar, and the shoreline is receding. Such a beach profile is called a storm profile or bar profile. In summer, the sea is relatively calm and the waves are small. The sediment accumulated in the offshore area is gradually pushed to the shore by the waves to form the beach shoulder. The increase of the beach shoulder makes the shoreline move forward to the seawall. Such a section is called the constant wave section or the beach shoulder section. Therefore, the changes of beach profile and shoreline caused by the lateral movement of sediment under the action of waves are short-term changes with an annual cycle.

1.2.4 Alongshore current

When the waves strikes the sandy coast with an oblique angle, the coastal current caused by the waves in the breaking zone carries the sediment along the coast longitudinally, which is an important reason for shaping the coastal landform and making the shoreline advance or retreat continuously. In general, if the coastal sediment transport rate entering a certain section of coast is greater than that exiting from the section of coast, the coastal sediment will accumulate and the shoreline will move forward. If the output is greater than the input, the coast is eroded and the shoreline recedes. If the sediment transport rates in two different directions are equal, that is, the net sediment transport rate is zero, or if the net sediment transport rate is not zero, but the net sediment transport rates in two adjacent sections are equal, then the coast is in dynamic equilibrium and there is no siltation or erosion along the shoreline.

1.2.5 Tide

In addition to wind and waves, tidal current is an important dynamic factor for smooth silty coast. Wind waves generally lift fine particles of sediment into suspension, while tidal currents cause large-scale horizontal transport of sediment. The Abandoned Yellow Estuary Delta in northern Jiangsu Province collapses constantly under the waves, the shoreline recedes, and the sediment is gradually transported along the coast to the south and north by the tidal current.

Because of the specialty of northern Jiangsu coastal tidal environment, the flood and ebb tide shows convergence and divergence phenomenon with Qiang Port as the center. The maximum tidal range of the eastern tidal channel south of the Abandoned Yellow River Estuary is 5.5 m, and the maximum tidal current velocity is more than 2 m/s. Strong trend still keep sediment along the Xiyang channel to Qinggang, making the Xiyang channel a typical silting and swelling section.

1.2.6 Wind

Wind is also an important motivator for modifying a beach. The wind with a large speed can

1.2 Variation influencing factors of coastal erosion and siltation

carry exposed sand particles from a sandy beach in three different ways. The smaller and lighter particles are thrown into the air stream for a distance, which is called suspension. The larger grains of sand are carried by the wind along the beach in a series of leaps, which is called saltation. If the particles roll or bounce along the beach surface under the action of wind, it is called surface fore hummocks. This natural transport results in the sorting of particles from the beach surface. Finer particles are removed from the beach surface, and the intermediate-sized sand particles form the fore hummocks, which are the dunes immediately following the back beach. The coarser particles remain at the original beach surface. General surface peristaltic particles account for about 20%-25% of the total number of particles transported, most particles are transported in a continuous jumping way. If there is an obstacle on the beach that reduces the wind speed and causes the sand particles to fall and accumulate, the formation of sand dunes begins. The gradual formation of dunes obstructs the landward transport of wind-carried sand and further accumulates nearby. The vegetation on the beach is often the trigger for the formation of forehills. Dune plays an important role in the protection and stability of wind-eroded beach.

1.2.7 Human activities

Human activities have different degrees of influence on the coastal wave, current and sediment migration, resulting in coast surface change, coast line adjustment or coast erosion and silt imbalance, which is mainly manifested in:

(1) Coastal engineering interfere with the natural motion of coastal sediment movement or other forms of sediment migration. Most commonly, port facilities or large structures interrupt the sediment supply from the shore downstream, causing erosion of its downstream neighbors. For example, the original drift sand from north to south along the coast of Friendship Port, Mauritania, was built in 1986, which was constructed by our country. After the port was built, the pier intercepted the sediment transport along the coast, resulting in erosion on the south coast of the port area and endangering the safety of land buildings. Another example is Skagen Harbor in the north of Jutland Peninsula, Denmark. The double jetty encirclements form a good harbor area, but at the same time, it also acts as a complete barrier to the coastal sediment, which causes serious flooding on the downstream side. Even if the groin dam is built on the downstream side, it can only slow down but not prevent the erosion along the shoreline due to the lack of sand supply.

(2) The tidal and reclamation project is conducive to land reclamation, cultivation and breeding, and the development of new industrial development zones. But at the same time, reclamation projects have significantly changed the flow field of the local sea area. If the planning is not proper, it may not only destroy the balance of erosion and sedimentation of nearby shoals, leading to the change of the original geomorphic form and sediment distribution, but also deteriorate the local ecological environment, leading to the decline or even extinction of some aquatic resources. In particular, the reclamation in the bay or the top of the tidal channel will make tidal prism of the bay and tidal area of tidal channel decrease significantly, resulting in channel silt-

ation or tidal channel atrophy. For example, in Jiaozhou Wan, Shandong Province, the bay head has been continuously reclaimed, which has reduced the bay area by 120 km^2 in 50 years. The decrease of tide capacity makes the bay mouth velocity and water crossing section smaller, leading to the Cangkou Channel, one of the main waterway of Qingdao Port, shrinking and moving westward.

(3) The construction of tidal barrier at the estuary of a river is conducive to tide and salt prevention and irrigation. At the same time, the construction of sluices in estuaries will cut off the runoff and sediment transport, resulting in the formation of upstream channel siltation. What is more serious is that the construction of tide gate makes the tide wave deformation intensified and the original moving wave becomes vertical wave. The duration of the flood tide is shortened, the flow velocity of the flood tide is increased, and the sediment from the sea is more silted downstream of the sluice. For example, the sluice has been built in almost all the estuaries along the northern Jiangsu coast, which truncates the channel of sea-river combined transport and makes it more difficult to build large ports in the estuaries. As a result, the northern Jiangsu coast has become one of the few port blank belts in China.

(4) Large-scale underground exploitation of oil, water and gas in the coastal zone may cause the subsidence of the land. Unplanned sand and stone mining on the beach cause erosion of the shoreline. The destruction of coastal and beach vegetation reduces the ability to withstand storm surges. Many gravel deposits along the coast of Liaoning Province were originally natural embankments along the coastline, such as the Bailanzi gravel embankment in Lvshun, which was provided by the ancient flood deposits in the Laotieshan area and was carried horizontally by waves. However, due to years of excessive mining, the shoreline has been forced to land continuously, and the houses on the top of the embankment have been forced to relocate three times since liberation. Xinjinpikou coast, since the 1970s a large number of sand and stone excavation, the results of sea erosion cliff in the last 20 years, retreat 0.5 to 1.0 m/a. Mackerel Wan in Gai County on the west coast of the Jiangdong Bandao, the original large beach, because of the construction of a large number of Yingkou Port mining beach sand, after 1969, the shoreline retreat 2 m/a, the coastal road constantly diverted or interrupted.

Exercise

Q1 Vocabulary explanation: Coastline, coastal zone and coast.
Q2 Fill in the blanks: Coastal zone include (   ), (   ), (   ), (   ), (   ) and (   ).
Q3 Fill in the blanks: The type of coastal zones include (   ), (   ), (   ), (   ) and coral reef coast.
Q4 Fill in the blanks: The morphological changes of the coastline and the evolution of the landform of the coastal zone are the result of the dynamic action of the sea and land,

and the important factors affecting the natural environment of the coastal zone are as follows: ( ), ( ), ( ), ( ).

Q5 Short answer: What are the environmental characteristics of national coastal zone?

Q6 Fill in the blanks: Common types of coastal zone resources include: ( ), ( ), ( ).

Q7 Short answer: What are the factors affecting the change of coastal erosion and siltation?

Q8 Short answer: What are the distribution and characteristics of Chinese muddy silt coast?

Chapter 2 Design Standards of Coastal Engineering Structures

The design standard of coastal engineering buildings mainly introduces the building grade and design standard, and calculates the design tidal level and wave based on this, and integrates the content of engineering hydrology into the design of coastal engineering buildings.

2.1 Design standards

2.1.1 Design standards for coastal protection buildings

Seawall design standard is the tide and wave combination standard based on which seawall can withstand and defend safely. Seawall design standards should first be determined by dividing seawalls into different grades according to the importance of the objects to be protected and the population or land area to be protected. The corresponding defense standards are stipulated according to different levels, then the tide level and wave value stabilized from a certain cumulative frequency and return period determined by the defense criteria are calculated.

Currently, there is no unified seawall design standard. In the 1960s and 1970s, Shanghai, Jiangsu, Zhejiang, Fujian, Guangdong and other coastal areas developed standards that were suitable for their specific conditions. These standards were not the same, but they were all reflected by a combination of characteristic tidal levels and typhoon winds. The following is a list of seawall defense standards in some areas for reference:

1. Guangdong Province (1968)

This is shown in the following Table 2.1.1.

Table 2.1.1   Guangdong Province

Seawall grade	Protection area/mu	Defense standards
1	>50,000	Surge level and category 10 typhoon
2	10,000 – 50,000	Surge level and category 9 typhoon
3	1,000 – 10,000	Surge level and category 8 typhoon
4	<1,000	Surge level

Note: 1 mu ≈ 0.667 hm^2.

2.1 Design standards

2. Fujian Province (1964)

This is shown in the following Table 2.1.2.

Table 2.1.2   **Fujian Province**

Seawall grade	Protection area/mu	Defense standard
1	>50,000	Historical high tidal level and category 12 typhoon
2	10,000 – 50,000	Historical high tidal level and category 11 typhoon
3	1,000 – 10,000	Historical high tidal level and category 10 typhoon
4	<1,000	Historical high tidal level and category 9 typhoon

3. Shanghai Municipality

The defense standards are: Historical high tidal level and category 11 typhoon.

4. Jiangsu Province

The defense standards are: Historical high tidal level and category 10 typhoon.

The earlier seawall defense standards listed above had several shortcomings:

(1) Using wind scale to reflect wave size is an indirect method to determine wave elements. Compared with using wave observation data directly, the error is larger. The standard around the style is not clear enough, and the necessary conversion and correction is lack.

(2) There is no clear frequency concept for the historical high tidal level or the highest historical tidal level (also known as surge level) used in the local standards. The return period represented by the highest tides in history vary greatly from place to place, so it has great arbitrariness and randomness. It is obviously inappropriate to use the highest tidal level in history as a design standard.

(3) Some historical high tidal levels or highest tidal levels are obtained through survey, and their reliability is often affected by the specific local conditions and survey and analysis methods. In some cases, for example, the tidal level may include wind increase and wave rise. The wind increase is related to the direction of the wind at the time. Wind rise is not only related to wind speed and direction, but also related to local topography, building type and so on. The survey data from different places show that with the increase of the number of years of the survey data, the value of the historical highest tidal level will vary greatly, and the data from the survey also has this problem, and the value is more unreliable.

In order to overcome the above shortcomings, the annual frequency statistics method has been adopted in the design standard of seawall since the late 1970s. In the sites with more than 20 years of continuous tidal level data, the highest annual tidal level was used as the sample for frequency analysis, and the high tidal levels in different return periods were calculated as the design high tidal levels. This allows data from different years of observation to be unified to a common standard, and the tidal levels identified have clear statistical implications.

Chapter 2 Design Standards of Coastal Engineering Structures

Design wave includes two aspects in the design standard. First, the recurrence interval standard of design wave refers to the average number of years that the wave of a particular wave train appears, and represents the long-term (decades or centuries) statistical distribution law of wave elements. The other is the standard of the cumulative frequency of wave train, which refers to the occurrence probability of a certain wave element in the actual irregular wave train on the sea surface, and it represents the short-term (tens of minutes) statistical distribution of wave elements.

Design tidal level return period and design wave return period are adopted as design standards in *Technical Regulations of Zhejiang Seawall Engineering*, as shown in Table 2.1.3.

Table 2.1.3   Project grade determination and design return period

Grade	Protection scope and importance	Design return period/a
Grade Ⅰ	The protected area is large and there is a significant impact on the national economy after the accident. Or the protected area is small, and there are significant industrial facilities in the protected area	50–100
Grade Ⅱ	The protected area is more than 50,000 mu or the population is more than 50,000 people	20–50
Grade Ⅲ	The protection area is between 10,000 mu and 50,000 mu, or the population is between 10,000 and 50,000 people	10–20
Grade Ⅳ	The protected area is less than 10,000 mu, or the population is less than 10,000 people	10

At the same time, the design attention period of the wave is the same as the return period of the design high tidal level. In Table 2.1.3, the design return period of seawall at all levels has upper and lower limits. If there are significant industrial and mining enterprises within the protected area, or if although the number of acres under protection is relatively small in island areas, but land is scarce and the population density is high, thus making their importance greater, the upper limit value can be adopted, or the level can be raised by one grade. If there is a submersible dam or silting works in front of the pond in the harbour area, which has a significant effect on wind and wave reduction, a level or the lower limit can be lowered.

In the *Technical Regulations for Reclamation Engineering Design of Fujian Province*, the classification of hydraulic buildings is first stipulated, as shown in Table 2.1.4.

Table 2.1.4   Classification of building levels

Project grade	Permanent building class		Temporary structure class
	Main building	Secondary building	
Large	3	4	5
Medium	4	5	5
Small	5	5	

2.1 Design standards

It also points out that the seawall is a permanent main structure. Then, two design tidal levels are specified. Table 2.1.5 shows that the design tidal level adopts the highest tidal level in history or the highest average annual tidal level. The design tidal level is determined by frequency analysis method in Table 2.1.6.

Table 2.1.5   Building grade and design tidal level, design wind and wave rise

Building grade	Design tidal level	Design wind		Wave rise accumulation rate/%	Security elevation/m
		North, Northeast	Other		
3	Recorded high tidal level	12	11	10	0.7
4	Recorded high tidal level	11	10	20	0.5
5	Mean annual maximum tidal level	10	10	30	0.3

Table 2.1.6   Building grade and design tidal level return period, design wind speed return period, wave rise accumulation rate and safe elevation

Building grade	Design tidal level return period/a	Design wind speed return period/a	Wave rise accumulation rate/%	Safe elevation/m
3	100 – 50	50	2	0.7
4	50 – 30	30	5	0.5
5	30 – 20	10	13	0.3

Calculate the top elevations of the seawall respectively according to the two different values of design tidal level and design wind condition listed in Table 2.1.5 and Table 2.1.6, and then compare with each other to find the largest one.

The return period standard mainly reflects the service life and importance of seawalls, while the cumulative frequency standard mainly reflects the different nature of tidal level or wave action on different types of seawalls or different parts. For example, for seawalls with large protected areas, the return period of tidal level and wave disturbances is longer. The vertical seawall is sensitive to wave action, and individual large waves in the wave train may affect the safety of the rebuilt object. Therefore, the repair of this kind of seawall is difficult, so the standard of wave accumulation frequency is adopted. The sloping seawall uses a lower cumulative frequency standard because individual large waves in the wave train do not play a decisive role in it, and once damaged, it is easier to repair. In addition, the cumulative frequency standards for different parts of the same type of building are also different, and the cumulative frequency standards should be designed to be higher for critical parts whose damage will affect the safety of the entire building.

The following factors should also be considered when deciding the design standard of seawalls:

(1) In the construction of seawall, there is a case that after several years of construction, the bund is silted up and the original seawall is sidelined, or the conditions of the original sea-

Chapter 2　Design Standards of Coastal Engineering Structures

wall will be greatly changed. These seawalls are called transitional seawalls. Obviously, this kind of seawall should have different design standards and different return periods to reflect the difference between this kind of seawall and the permanent seawall with no significant change in long-term conditions.

(2) Different seawall types with different sensitivity to damage caused by overtopping should also have different design standards. A seawall primarily constructed with fill soil in a sloping style, once overtopped by wind and waves, is prone to erosion of the embankment body, which easily leads to collapse and damage, indicating a poor ability to resist destruction caused by wave overtopping. The straight wall seawall, with masonry as the main body, can maintain a certain stability and not collapse even if the water washes away all the soil attached to the embankment. For these two types of seawalls with different materials and structures, different wave accumulation frequency should be adopted to reflect the difference.

(3) Different standards should correspond to different durations of wave action on the seawall. The duration of wave action refers to the sustained period during the design recurrence interval where the average wind speed is above force 6, and during each high tide process, the tidal level remains above the warning water level for more than 2 hours. Since the duration of designed wave action reflects the length of time overtopping may occur and the amount of overtopping, different wave series cumulative frequencies are also used to reflect the differences in design standards.

The standard of wave train cumulative frequency suggested by the *Technical Regulations of Seawall Engineering in Zhejiang Province* is shown in Table 2.1.7.

Table 2.1.7　　　　　　　　**Wave train cumulative frequency**

Seawall type	Position	Calculation content	Wave train cumulative frequency
Vertical type	Crest level	Wave climb calculation	2% (overtopping waves are not allowed)
			13% (overtopping waves are allowed partly)
	Wave wall, wall body, gate, gate wall	Strength and stability	1%
	Base bed, bottom block	Stability	5%
Slope type	Pool top (elevation)	Wave climb calculation	2% (overtopping waves are not allowed)
			13% (overtopping waves are allowed partly)
	Wave wall, gate, gate wall	Strength and stability	1%
	Armour rock, bottom block	Stability	13%

2.1 Design standards

2.1.2 Design standards for breakwaters

The design condition of breakwater includes natural condition, service condition, material condition and construction condition.

Natural conditions include wind, wave, tidal level, ocean current, sediment, ice and topography, geomorphology, geology, earthquake and other conditions. In addition, water quality and marine life should be investigated.

Service conditions mainly reflect the specific requirements of the port for breakwater. For example, the requirements of the breakwater diffraction, reflection (including side reflection), wave penetration, wave crossing, additional requirements in sand prevention, flow prevention and ice prevention, the use of the top of the embankment as a channel or the erection of the pipeline, the requirements of lighting or lighthouse set on the top of the embankment, the inside of the embankment and as a pier requirements. In the design of breakwater, it is necessary to consider various loads and forces caused by different requirements on use.

In terms of material conditions, the most basic is the supply of bulk sand and stone. The quality of stone, the maximum weight of large blocks of stone which may be mined and transported, the possibility of using strip stone and the price of various materials affect the cost of breakwater.

Construction conditions include the existing large construction facilities (such as precast concrete components, caisson slideway, etc.), the conditions of the construction machinery equipment, the possible construction technology, the local construction days and the allowable construction period, etc.

In the comparison and selection of breakwater projects, the reasonable structure type should be determined by the comprehensive comparison of technology and economy according to the specific local design conditions, combined with the characteristics and applicable conditions of various breakwaters.

It should be pointed out that since the axial length, water depth, wave and soil condition of the foundation may vary along the breakwater, it is economical and reasonable to divide the whole length of the breakwater appropriately and then adopt different structural types or different section scales according to the specific conditions.

For example, in the east breakwater of Shanhaiguan Shipyard, the sloping embankment with concrete artificial block face is adopted in the shallow water section within the -6.5 m isobath, and the vertical embankment with gravity structure is adopted beyond the -6.5 m isobath. In all kinds of natural conditions, the wave has special and important significance to the design of breakwater. Therefore, the following will make an in-depth discussion on the design of the wave standard of breakwater.

1. Designed wave standard

In determining the force of wave on various types of coastal engineering structures, it is necessary to define a reasonable and representative wave element, namely, design wave. Because the waves on the sea are changing at any time, so the design of waves should have a certain statistical significance. In general, the criteria for designing a wave include: designing the return period of the wave and designing the cumulative frequency of the wave train. The cumulative frequency of the design wave train refers to the occurrence probability of the design wave element in the irregular wave train, which represents the short-term (measured in tens of minutes) statistical distribution rule of the wave element. During this statistical period, we consider the sea surface is in a steady state, that is, the average state of wave elements does not change with time. The return period of designed waves refers to how many years the average wave of a particular wave train with cumulative frequency appears, which represents the long-term (decades) statistical distribution law of wave elements.

The two aspects of designing wave standards have different meanings and should be specified separately considering different factors. The standard of return period of design wave mainly reflects the service years and importance of the building, while the standard of cumulative frequency of design wave mainly reflects the different effects of waves on different types of marine structures.

In the People's Republic of China industry standard *Port and Waterway Hydrological Code* (JTJ 213—98) there are clear provisions for the design standards of all kinds of buildings in port engineering, which can be used as reference for general coastal engineering buildings.

It is stipulated in the *Code of Hydrology for Harbour and Waterways* (JTS 145—2015) that the return period of the design wave should be 50 years when calculating the strength and stability of straight wall type, pier type, pile type and general sloping type structures. For non-important buildings such as sloping revetment, which will not cause significant loss after destruction, the design wave return period can be adopted for 25 years. For buildings of special importance, such as sea lighthouses, when the measured wave height is higher than the wave height of the cumulative frequency of the same wave train with a return period of 50 years, the standard can be appropriately raised, and if necessary, it can be calculated using the measured wave height.

It is stipulated in the *Code of Hydrology for Harbour and Waterways* (JTS 145—2015) that when calculating the strength and stability of straight wall, pier, pile and sloping buildings, the standard of wave train cumulative frequency for the design wave height shall be adopted according to Table 2.1.8.

When the calculated wave height is higher than the shallow limit wave height, the limit wave height should be used. The average wave period can be adopted, and the wavelength can be calculated as follows:

2.1 Design standards

$$L = \frac{g\overline{T}^2}{2\pi} \tanh \frac{2\pi d}{L} \tag{2.1.1}$$

Where, L is wave length, m; $\overline{T}$ is average period, s; g is gravitational acceleration, m/s²; d is depth of water, m.

Table 2.1.8　　Cumulative frequency standard for designing wave height

Building type	Position	Calculation content	Wave height cumulative frequency/%
Straight wall type, pier type	Superstructure, wall body, pile foundation	Strength and stability	1
	Base bed, bottom block	Stability	5
Sloping type	Parapet or dike square	Strength and stability	1
	Protective block stone, protective block	Stability	13
	Bottom block	Stability	13

Note: When the ratio of mean wave height to water depth is less than $\frac{\overline{H}}{d} < 0.3$, $F = 5\%$ is recommended.

2. The return period of design waves

According to the *Code of Hydrology for Harbour and Waterways* (JTS 145—2015), design wave heights for different return periods can be determined under the following three conditions:

(1) When long-term measured wave data are available at or near the project location, the annual maximum series of a certain cumulative frequency wave height in the sub-direction can be used for frequency analysis.

(2) When there is no long-term wave measurement data at or near the project location, but the distance between the opposite bank is less than 100 km, frequency analysis is carried out by using the annual maximum series of wind speed in different directions, and then the design wave height of the same return period is indirectly determined by the wind speed value and the distance between the opposite bank in a certain return period.

(3) When there is no long-term wave data and the distance between the opposite bank is long, the historical weather chart can be used to select the most unfavorable weather process in each direction each year, and the annual maximum wave height can be calculated by the relevant method, and then the frequency analysis can be carried out.

The wave period corresponding to the design wave height of a return period (mean period) is calculated as follows:

1) When large local waves are mainly wind waves, the corresponding period of the design wave height can be extrapolated from the correlation between wave height and wave period of local waves, or the corresponding effective wave period T_S can be determined first according to Table 2.1.9, and then the corresponding mean period $\overline{T}$ can be obtained by $T_S = 1.15\overline{T}$.

Chapter 2　Design Standards of Coastal Engineering Structures

Table 2.1.9　Approximate relation between wave height and period of wind wave

$H_{13\%}/m$	2	3	4	5	6	7	8	9	10
T_S/s	6.1	7.5	8.7	9.8	10.6	11.4	12.1	12.7	13.2

2) When the large local waves are mainly swell or mixed waves, the frequency analysis can be carried out by using the cycle series corresponding to the maximum annual wave height, so as to determine the cycle value of the same return period as the designed wave height.

For frequency analysis of wave height or period, the continuous data should not be less than 20 years old.

The frequency curve of wave height and period is generally Pearson Type Ⅲ curve. When conditions permit, the principle of best fit with actual measurement data can be applied, selecting other theoretical frequency curves such as the Type Ⅰ Extreme Value distribution, log-normal distribution, and Weibull distribution, etc., ultimately determining the design waves for different return periods.

According to the statistical results of eight ports in the north and south of China, the ratio of 50-year wave height to 25-year wave height and 100-year wave height to 50-year wave height is about 1.1.

It should be noted that the return period is an average concept, so if the building is designed with 50-year waves, there is no guarantee that there will be no more than 50-year waves during the life of the building.

According to probability theory, the probability that a wave with a return period of T_R will occur in n years is:

$$q = 1 - \left(1 - \frac{1}{T_R}\right)^n \quad (2.1.2)$$

Where, q is risk rate or probability of encounter, q is the value for different T_R; n is given in Table 2.1.10 according to Eq. (2.1.2).

Table 2.1.10　　　　　　　　　　　Risk rate q

n	q					
	$T_R=5$	$T_R=10$	$T_R=20$	$T_R=25$	$T_R=30$	$T_R=50$
5	0.672	0.410	0.226	0.185	0.156	0.096
10	0.893	0.651	0.401	0.335	0.288	0.183
20	0.988	0.878	0.642	0.558	0.492	0.332
25	0.996	0.928	0.723	0.640	0.572	0.397
30	0.999	0.958	0.785	0.706	0.638	0.455
50	0.999	0.995	0.923	0.870	0.816	0.636

As can be seen from Table 2.1.10, if the service years of the breakwater are $n=50$, and $T_R=50$ is also taken as the recurrence period, the probability of the waves greater than or equal to the design wave reaches 63.6%. Conversely, if $n = 50$ and $q = 5\%$, $T_R = 975$ a can be calculated from Eq. (2.1.2). In the report of the 22nd International Conference on Shipping, France had proposed a relatively high standard of using once-in-a-thousand year design waves for straight wall and pile structures, which was estimated at a risk rate of 5%. Of course, such a high standard, the actual engineering design is not used, the reasons can be considered as follows:

1) Any project design, must balance between the two factors of safety and economy. If designed standards are too high, will inevitably increase the cost of the project. In the final analysis, for offshore buildings, it is often uneconomical and even difficult to achieve absolute safety.

2) The risk of designed waves is not the same as the risk of building accidents. Because the general building has a certain degree of safety in the case of wave design, so if the wave slightly exceeds, it will not cause significant damage.

In Part I of the *British Code for Offshore Structures* (BS 6349), it is noted that when using the general method of extrapolating frequency curves to estimate waves with recurrence in tervals larger than approximately a 100 - year event. It is necessary to note that their reliability will be affected by long-term changes in climate conditions. In addition, the frequency curve may be affected by water depth. In general, it can be considered that there is no method to estimate wave elements when the recurrence period is very long.

2.2 Design tidal level calculation

2.2.1 Design high tidal level calculation

The design high tidal level is usually determined by calculating the frequency of annual maximum tidal level. In the semi-diurnal tide area, there are more than 700 high water data in the annual tidal level observation, which are produced under the comprehensive action of astronomical and meteorological factors. So it has some random properties. The number of high tides in a year can be approximated as a series of random variables in which the largest term is the value of the annual high tides. If one have n years of tide data, one have n maximum terms. Since the largest term is the extreme value, its probability distribution is called the extreme value distribution.

In the highest tidal level of n years, the frequency of occurrence of a tidal level value greater than or equal to P (%), and the return period T (year) corresponding to this value is:

$$T = \frac{100}{P} \tag{2.2.1}$$

The return period is an average concept. The so-called return period of 50-year tides does not happen exactly once every 50 years. In fact, it may happen a few times in

50 years, or it may not happen at all. It merely indicates that, on average, over a long period of time, it occurs once every 50 years. Let $T=h_{50}$ of a 50-year tide, so $P=100/T=2\%$, It represents a 2% chance of tides greater than or equal to h_{50} occurring each year, and a $1-P$ chance of no h_{50} occurring each year. Assuming that the designed service life of the seawall is N years, the probability that the tidal level is less than h_{50} during its service life is:

$$F=(1-P)^N \qquad (2.2.2)$$

Where, F is the safety rate.

And the risk rate q of encountering h_{50} during its service life is equal to:

$$q=1-(1-P)^N \qquad (2.2.3)$$

If the service life $N=50$ years, it can be calculated from the Eq. (2.2.2) and Eq. (2.2.3) respectively that within the service life of 50 years, the safety rate of sewall without encountering the once-in-50-year tide level h_{50} is only 36.4%, while the risk rate is 63.6%.

According to the regulations of our port engineering standards, the annual maximum tide frequency analysis requires continuous data of not less than 20 years, and the special tidal level should be investigated in the history. As for the line line of frequency analysis of the highest tidal level, the theoretical frequency curve of the extreme value Type I extreme distribution (Gumbel distribution) can be generally adopted, and the theoretical frequency curve of the Pearson Type III distribution can be adopted in the tidal sand estuary area affected by runoff.

The method of frequency analysis using extreme Type I is as follows:

Given n annual highest tide series h_i, its mean $\overline{h}$ is:

$$\overline{h}=\frac{1}{n}\sum_{i=1}^{n}h_i \qquad (2.2.4)$$

The mean square error S of the tidal level h_i in n years is:

$$S=\sqrt{\frac{1}{n}\sum_{i=1}^{n}h_i^2-(\overline{h})^2} \qquad (2.2.5)$$

The high water potential value h_p corresponding to the annual frequency P (%) is:

$$h_p=\overline{h}+\lambda S \qquad (2.2.6)$$

The coefficients λ related to annual frequency P (%) and data years n in Eq. (2.2.6) are shown in Table 2.2.1.

Table 2.2.1　　　　　Type I extreme value distribution law

n/a	Frequency $P/\%$											
	0.1	0.2	0.5	1.0	2.0	4.0	5.0	10.0	25.0	50.0	75.0	90.0
8	7.103	6.336	5.321	4.550	3.779	3.001	2.749	1.953	0.842	−0.130	−0.897	−1.458
9	6.909	6.162	5.174	4.425	3.673	2.916	2.670	1.895	0.814	−0.133	−0.879	−1.426
10	6.752	6.021	5.055	4.322	3.587	2.847	2.606	1.848	0.790	−0.136	−0.865	−1.400

2.2 Design tidal level calculation

Continued

n/a	Frequency P/%											
	0.1	0.2	0.5	1.0	2.0	4.0	5.0	10.0	25.0	50.0	75.0	90.0
11	6.622	5.905	4.957	4.238	3.516	2.789	2.553	1.809	0.771	−0.138	−0.854	−1.378
12	6.513	5.807	4.874	4.166	3.456	2.741	2.509	1.777	0.755	−0.139	−0.844	−1.360
13	6.418	5.723	4.802	4.105	3.404	2.699	2.470	1.748	0.741	−0.141	−0.836	−1.345
14	6.337	5.650	4.741	4.052	3.360	2.663	2.437	1.724	0.729	−0.142	−0.829	−1.331
15	6.266	5.586	4.687	4.005	3.321	2.632	2.408	1.703	0.718	−0.143	−0.823	−1.320
16	6.196	5.523	4.634	3.959	3.283	2.601	2.379	1.682	0.708	−0.145	−0.817	−1.308
17	6.137	5.471	4.589	3.921	3.250	2.575	2.355	1.664	0.699	−0.146	−0.811	−1.299
18	6.087	5.426	4.551	3.888	3.223	2.552	2.335	1.649	0.692	−0.146	−0.807	−1.291
19	6.043	5.387	4.518	3.860	3.199	2.533	2.317	1.636	0.685	−0.147	−0.803	−1.283
20	6.006	5.354	4.490	3.836	3.179	2.517	2.302	1.625	0.680	−0.148	−0.800	−1.277
22	5.933	5.288	4.435	3.788	3.138	2.484	2.272	1.603	0.669	−0.149	−0.794	−1.265
24	5.870	5.232	4.387	3.747	3.104	2.457	2.246	1.584	0.659	−0.150	−0.788	−1.255
26	5.816	5.183	4.346	3.711	3.074	2.433	2.224	1.568	0.651	−0.151	−0.783	−1.246
28	5.769	5.141	4.310	3.681	3.048	2.412	2.205	1.553	0.644	−0.152	−0.799	−1.239
30	5.727	5.104	4.279	3.653	3.026	2.393	2.188	1.541	0.638	−0.153	−0.776	−1.232
35	5.642	5.027	4.214	3.598	2.979	2.356	2.153	1.515	0.625	−0.154	−0.768	−1.218
40	5.576	4.968	4.164	3.554	2.942	2.326	2.126	1.495	0.615	−0.155	−0.762	−1.208
45	5.522	4.920	4.123	3.519	2.913	2.303	2.104	1.479	0.607	−0.156	−0.758	−1.198
50	5.479	4.881	4.090	3.491	2.889	2.283	2.086	1.466	0.601	−0.157	−0.754	−1.191
60	5.410	4.820	4.038	3.446	2.852	2.253	2.059	1.446	0.591	−0.158	−0.748	−1.180
70	5.359	4.774	4.000	3.413	2.824	2.230	2.038	1.430	0.583	−0.159	−0.744	−1.172
80	5.319	4.738	3.970	3.387	2.802	2.213	2.022	1.419	0.577	−0.159	−0.740	−1.165
90	5.287	4.709	3.945	3.366	2.784	2.199	2.008	1.409	0.572	−0.160	−0.737	−1.160
100	5.261	4.686	3.925	3.349	2.770	2.187	1.998	1.401	0.568	−0.160	−0.735	−1.155
200	5.130	4.568	3.826	3.263	2.698	2.129	1.944	1.362	0.549	−0.162	−0.723	−1.134
500	5.032	4.481	3.752	3.200	2.645	2.086	1.905	1.333	0.535	−0.164	−0.714	−1.117
1,000	4.992	4.445	3.722	3.174	2.623	2.069	1.889	1.321	0.529	−0.164	−0.710	−1.110
∞	4.936	4.395	3.679	3.137	2.592	2.044	1.886	1.305	0.520	−0.164	−0.705	−1.110

After calculating the values h_p corresponding to different P in Eq. (2.2.6), the theoretical frequency curve of the high tidal level can be given on the probability grid paper, and the empirical frequency points can be drawn at the same time to test the degree of cooperation. In the annual highest tidal level arranged in descending order, the empirical frequency P of the m-th item can be calculated as follows:

$$P=\frac{m}{n+1}\times 100\%\qquad(2.2.7)$$

The relation between the return period T (year) and the annual frequency P (%) is shown in Eq. (2.2.1).

If the Pearson Type Ⅲ distribution is used for high water frequency analysis, please refer to Section 2.3, *Design Wave Calculation* for the method of using the Pearson Type Ⅲ distribution to calculate design wave elements for different return periods. The only difference is that the annual maximum wave height sequence is replaced by the annual maximum tidal level sequence.

2.2.2 Design high tidal level calculation with very large values

If the survey results show that there has been an extremely high tidal level h_N in the past n years in addition to the original tide data, the frequency analysis should be carried out in the following way:

Its mean $\bar{h}$ is:

$$\bar{h}=\frac{1}{N}(h_N)+\frac{N-1}{n}\sum_{i=1}^{n}h_i\qquad(2.2.8)$$

Variance S should be changed to:

$$S=\sqrt{\frac{1}{N}\left(h_N^2+\frac{N-1}{n}\sum_{i=1}^{n}h_i^2\right)-(\bar{h})^2}\qquad(2.2.9)$$

At this time, h_p is still calculated by Eq. (2.2.6), but $\bar{h}$ and S in Formula should be calculated by Eq. (2.2.8) and Eq. (2.2.9) respectively. λ in formula is still found by Table 2.1.8, but the data years are equal to N instead of n.

The experience point of extra-high tidal level is still calculated by Eq. (2.2.7).

2.2.3 Design high tidal level calculation with data shortage

The design high tidal level can be deduced by "extreme synchronous difference ratio method" when the long-term observation data of tidal level is the least in the proposed construction area. The condition is that the proposed project area should have more than 5 years of observation data as the fixed point. As a known point with no less than 20 years of continuous observation data, it should meet the conditions of similar tidal sand properties, similar geographical location, similar influence of river runoff (including flood season) and similar influence of water increase and decrease between it and the fixed point. It is then assumed that the difference between the design high tidal level and mean sea level for the same return period in both places is the same as the difference between the mean annual high tidal level and mean sea level in both places as λ of the Type Ⅰ extreme distribution law in Table 2.1.8.

$$\frac{h_y-A_y}{h_x-A_x}=\frac{R_y}{R_x}\qquad(2.2.10)$$

Where, h_x and h_y are the same design high water position in the recurrence period of the

2.2 Design tidal level calculation

constant point and the pending point respectively; A_x and A_y are the mean sea level of known points and specific points respectively; R_x and R_y are the difference between the mean annual highest tidal level and the mean sea level of known points and fixed points respectively.

$$h_y = A_y + \frac{R_y}{R_x}(h_x - A_x) \qquad (2.2.11)$$

If the proposed construction site lacks even five years of continuous tidal observation data, at this point, short-term synchronized tidal level observations should be conducted at the site (the target location) and in nearby regions with available data (known points). For example, a synchronized observation for one month could be performed, following which the design high tide level for the target location can also be estimated. The method is as follows:

(1) The short-term mean sea level of the two places in the period of high tidal range in the middle of the month is calculated by using the data of one month observed synchronously, that is, the arithmetic mean value of measured tidal level in this period. A_{xx} and A_{yy}, respectively. Subscripts x and y represent known points and pending points (same below).

(2) Using the tidal level data of this month, the tidal level processes of the two places are plotted respectively, and the mean sea level is overlapped, and the high tide and low tide time of the two process lines are as consistent as possible, comparing whether the tide type and tidal range are similar.

(3) If the two tides are similar, the difference in tidal level is caused by the difference in tidal range between the two places. Therefore, it can be assumed that the difference between the design high tidal levels h_x and h_y and the multi-year mean sea level A_x and A_y during the same return period in both places is proportional to the tidal range in both places, namely:

$$\frac{h_y - A_y}{h_x - A_x} = \frac{R_y}{R_x} \qquad (2.2.12)$$

Although the formula is in the same form as the Eq. (2.2.10), the R_y/R_x on the right side of the equal sign here represents the mean value of the tidal range ratio between the two places, that is, the tidal range R_{yi} and R_{xi} of the two places are calculated respectively twice a day, and $R_{yi}/R_{xi} = (R_y/R_x)_i$, then calculate $R_y/R_x = [(N-1)/n] \sum_{i=1}^{n}(R_y/R_x)_i$, Here n is the actual count of the tidal range in a month.

(4) Assume that the difference between the short-term mean sea level and the multi-year mean sea level is the same, i.e.:

$$\left.\begin{array}{l} A_y - A_{sy} = A_x - A_{sx} = A \\ A_y = A_{sy} + (A_x - A_{sx}) \end{array}\right\} \qquad (2.2.13)$$

Where, A is the monthly correction of sea level in the two places.

(5) Calculation.

$$h_y = \frac{R_y}{R_x}(h_x - A_x) + A_{sy} + (A_x - A_{sx}) \qquad (2.2.14)$$

Where, A_x is the average sea level at the known point for many years; h_x is the annual design high tidal level at a certain recurrence period at a known point; h_y is the design high tidal level at a fixed point with the same recurrence period as h_x.

Fig. 2.2.1 shows the relationship of characteristic values of each tidal level in Eq. (2.2.13) and Eq. (2.2.14).

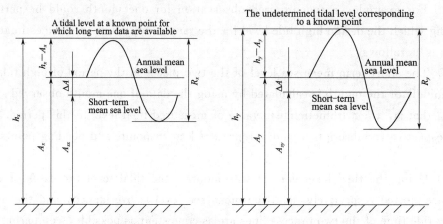

Fig. 2.2.1 Relationship between characteristic values of tidal level at two stations

2.3 Design wave calculation

As mentioned in the previous chapter, design wave standard include recurrence period standard and wave train cumulative frequency standard. The following are the two design wave calculation methods.

2.3.1 Wave height cumulative frequency calculation

The cumulative frequency of wave train includes the cumulative frequency of wave height and period. For wave height, the opportunity for waves equal to or greater than a certain wave height to appear in a wave series is known as the cumulative frequency of that wave height.

(1) According to the wave search data measured by the wave measuring instrument at fixed point, the wave height values of 100 – 150 consecutive waves are read by the upper span zero method. In any wave series, wave height is a random variable in the random event of the vertical displacement of the wave surface. We take all the wave heights of a wind wave in a steady state process as the population, then a continuous wave height record taken from it is the sample.

2.3 Design wave calculation

Table 2.3.1 is a series of examples of 100 waves continuously observed at a wave observation station in China.

Table 2.3.1 Actual continuous observation data of a wave observation station

H/m	T/s	H/m	T/s	H/m	T/s	H/m	T/s	H/m	T/s
2.0	9.2	1.3	5.3	0.8	4.5	0.6	11.4	2.1	9.2
3.0	6.6	3.2	7.3	2.5	6.6	1.4	6.6	2.7	9.8
2.5	6.6	5.3	6.8	4.1	7.3	1.6	5.6	3.2	8.6
3.1	6.9	3.3	6.9	3.8	6.9	1.1	5.3	1.9	5.6
1.6	8.6	1.5	8.3	1.7	6.9	1.6	8.3	0.2	4.1
1.9	7.1	1.2	8.6	1.0	5.3	2.1	6.0	1.4	7.9
2.2	5.4	1.9	6.6	2.0	5.8	1.1	23.0	2.1	5.6
3.3	7.1	1.2	8.6	1.0	5.3	2.1	6.0	1.4	7.9
3.0	6.6	3.1	6.6	2.0	9.4	2.6	6.9	2.2	7.9
4.9	7.5	1.8	6.4	1.8	8.3	1.7	8.8	2.1	6.4
1.6	8.1	1.4	4.5	1.3	9.6	1.5	4.5	1.6	7.5
1.5	8.1	1.8	5.8	1.3	6.8	3.9	7.1	1.3	8.3
0.9	4.3	1.8	6.2	1.5	5.4	3.0	8.1	2.4	7.5
1.1	5.4	1.5	4.3	1.0	4.1	2.4	16.1	3.7	7.3
3.1	7.5	4.3	6.6	2.0	5.8	3.3	6.2	3.8	6.4
3.2	6.8	4.8	7.1	1.4	7.5	2.0	6.4	2.4	6.2
2.3	6.6	4.1	6.9	0.3	3.6	1.1	6.2	2.6	7.3
1.2	4.5	3.9	6.6	1.3	10.5	2.5	5.8	1.3	4.3
1.5	4.9	2.9	6.4	2.0	8.4	2.1	5.3	2.2	6.8
2.7	6.2	0.7	4.1	2.0	8.1	3.5	7.1	3.3	8.1

(2) By sorting the wave height H from large to small, there are m wave heights of different sizes. By calculating the frequency n_i of each wave height, the average wave height $\overline{H}$ of the wave series can be calculated as follows:

$$\overline{H} = \frac{1}{N}\sum_{i=0}^{n} n_i H_i \qquad (2.3.1)$$

$$N = \sum_{i=0}^{n} n_i$$

Where, N is the total number of wave heights taken in the series, in the case, $N=100$, $m=38$, then $\overline{H}=2.2$ m.

(3) Calculate relative wave height $\dfrac{H_i}{\overline{H}}$, called the modulus ratio coefficient K. The modulus ratio coefficient of average wave height is equal to 1.0.

(4) The series are divided into several groups with appropriate group distance $\dfrac{\Delta H}{\overline{H}}$,

and the corresponding wave heights of the upper and lower limits of each group are calculated. In this example, the group distance is $\frac{\Delta H}{\overline{H}} = 0.2$, as shown in columns 1 and 2 of Table 2.3.2.

Table 2.3.2   Statistical table of cumulative frequency of wave train

$\frac{H_i}{\overline{H}}$	H	$\frac{\Delta H}{\overline{H}}$	n_i	$f_i = \frac{n_i}{N}$	$f_i / \frac{\Delta H}{\overline{H}}$	Σn_i	$F_i / \%$
1	2	3	4	5	6	7	8
2.4 - 2.2	5.3≥H>4.8		2	0.02	0.10	2	2
2.2 - 2.0	4.8≥H>4.4		1	0.01	0.05	3	3
2.0 - 1.8	4.4≥H>4.0		3	0.03	0.15	6	6
1.8 - 1.6	4.0≥H>3.5		5	0.05	0.25	11	11
1.6 - 1.4	3.5≥H>3.1		9	0.09	0.45	20	20
1.4 - 1.2	3.1≥H>2.6	0.2	10	0.10	0.50	30	30
1.2 - 1.0	2.6≥H>2.2		9	0.09	0.45	39	39
1.0 - 0.8	2.2≥H>1.8		18	0.18	0.90	57	57
0.8 - 0.6	1.8≥H>1.3		23	0.23	1.15	80	80
0.6 - 0.4	1.3≥H>0.9		14	0.14	0.70	94	94
0.4 - 0.2	0.9≥H>0.4		4	0.04	0.20	98	98
0.2 - 0.0	0.4≥H>0.0		2	0.02	0.10	100	100
			100	1.0			

(5) The frequency n_i of wave height in each group is recounted. Note the difference in the same Eq. (2.3.1). After the wave heights are grouped in Table n_i according to the upper and lower limits in column 2 of Table 2.3.2, the frequency of wave heights falling into the range of each group is listed in column 4 of the table.

(6) The frequency of each group is divided by the total frequency N to obtain the interval frequency of each group's wave height $f_i = \frac{n_i}{N}$. Obviously, $\Sigma f_i = 1.0$, The interval frequency reflects the frequency of occurrence of wave height in each group. Generally, the occurrence rate of wave height group close to the average wave height is larger, while the occurrence frequency of maximum and minimum wave height is smaller.

(7) Since the frequency of an interval f_i represents the total frequency of each wave height in the interval, assuming that the occurrence chances of any wave height in the interval are the same, $f_i / \frac{\Delta H}{\overline{H}}$ represents the average frequency of each wave height in the interval, which is listed in column 6 of Table 2.3.2.

With relative wave height $\frac{H_i}{\overline{H}}$ as the vertical coordinate and average frequency as the

2.3 Design wave calculation

horizontal coordinate, the average frequency histogram shown in Fig. 2.3.1 (a) can be drawn. In the figure, the division of the vertical coordinate is equal to the group distance $\frac{\Delta H}{\overline{H}}$, so the average frequency $f_i/(\Delta H/\overline{H})$ is high, and the rectangular area at the bottom of the group distance represents the interval frequency f of each group.

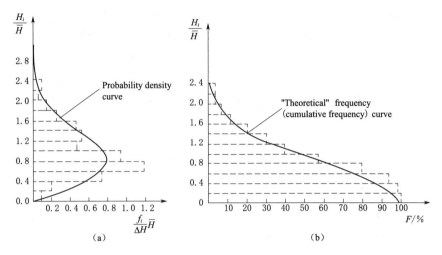

Fig. 2.3.1 Statistical graph of probability density and cumulative frequency

And the sum of all the rectangles on the histogram is equal to 1. When the group distance tends to the infinitesimal histogram, the height of the rectangle tends to form a curve, that is, the frequency density curve.

(8) In order to calculate the cumulative frequency of wave train whose height is greater than or equal to a certain wave height, the sum of cumulative times of wave height $\sum n_i$ greater than or equal to the lower limit of each group is calculated first, which is listed in column 7 of Table 2.3.2. Then $\sum n_i$ divided by the total degree N to find the cumulative frequency F_i greater than or equal to the height of a wave, as shown in column 8.

$$F_i = \frac{\sum n_i}{N} \times 100\% \qquad (2.3.2)$$

With F as the abscissa and relative wave height $\frac{H_i}{\overline{H}}$ as the ordinate, the cumulative experience frequency diagram can be drawn, as shown in Fig. 2.3.1 (b). When the interval size of grouping tends towards infinitely small, the height of the cumulative frequency graph rectangle tends to form a continuous frequency distribution curve, which corresponds to the cumulative frequency value F of any relative wave height, equal to the area of the frequency density curve and the vertical axis surrounding the relative wave height.

So, we find the statistical distribution law of wave height in wave train by statistical method. Generally, the statistical properties of waves can be fully reflected by selecting

100 – 150 consecutive waves for statistics. This is equivalent to the fluctuation duration of 10 – 20 min. If the wave number is too small, the representativeness of the sample is poor, which makes the statistical results unstable. If the wave number is too large, the wave cannot be guaranteed to be in a steady state, which affects the reliability of statistical results.

The statistical analysis of a large number of wave observation data shows that although the wave scales are varied and different, the frequency distribution characteristics of each wave train are very similar, and the wave distribution law can be deduced theoretically.

In deep water, wave height obeys Rayleigh distribution, and its distribution function is expressed as:

$$F(H) = \exp\left[-\frac{\pi}{4}\left(\frac{H}{\overline{H}}\right)^2\right] \qquad (2.3.3)$$

The curve in Fig. 2.3.1 (b) is the theoretical cumulative frequency curve of wave height. Eq. (2.3.3) can also be rewritten as:

$$\frac{H_F}{\overline{H}} = \left(\frac{4}{\pi}\ln\frac{1}{F}\right)^{\frac{1}{2}} \qquad (2.3.4)$$

That is, the relative wave height of a given cumulative frequency F is deduced.

In shallow water, the distribution law of wave height changes, and the wave height distribution can be written as:

$$F(H) = \exp\left[-\frac{\pi}{4\left(1+\frac{H}{d\sqrt{2\pi}}\right)}\left(\frac{H}{\overline{H}}\right)^{\frac{2}{1-\frac{H}{d}}}\right] \qquad (2.3.5)$$

It's called the Grukhovsky shallow water wave height distribution. Where d is the water depth.

Eq. (2.3.5) can also be written as:

$$\frac{H_F}{\overline{H}} = \left[\frac{4}{\pi}\left(1+\frac{H}{d\sqrt{2\pi}}\right)\ln\frac{1}{F}\right]^{\frac{1-\frac{H}{d}}{2}} \qquad (2.3.6)$$

When $\frac{H}{d}$ trends to 0, it is deep water, Eq. (2.3.6) automatically turns into Eq. (2.3.4). For convenience, the wave height mode ratio coefficients $\frac{H_F}{\overline{H}}$ corresponding to cumulative frequencies in different cases $\frac{H}{d}$ are calculated according to Eq. (2.3.4) and Eq. (2.3.6) and are listed in Table 2.3.3, H_F indicates the wave height of a certain cumulative frequency.

2.3 Design wave calculation

Table 2.3.3 Wave height mode ratio coefficients

$F/\%$	0.0	0.1	0.2	0.3	0.4	0.5
0.5	2.597	2.403	2.213	2.029	1.854	1.687
1.0	2.421	2.256	2.092	1.932	1.777	1.628
2.0	2.232	2.096	1.960	1.825	1.692	1.563
5.0	1.953	1.859	1.762	1.662	1.562	1.463
10.0	1.712	1.651	1.586	1.516	1.444	1.369
20.0	1.432	1.406	1.374	1.337	1.296	1.252
30.0	1.238	1.233	1.223	1.208	1.186	1.164
40.0	1.080	1.091	1.097	1.098	1.095	1.088
50.0	0.939	0.962	0.981	0.996	1.007	1.014
60.0	0.806	0.839	0.868	0.895	0.919	0.940
70.0	0.674	0.713	0.752	0.789	0.825	0.859
80.0	0.533	0.578	0.623	0.670	0.717	0.764
90.0	0.366	0.412	0.462	0.515	0.572	0.633
95.0	0.256	0.298	0.346	0.400	0.461	0.529

In the example, the deep water wave series listed in Table 2.3.3 has been calculated $\overline{H}=2.2$ m, and try to calculate wave heights $H_{1\%}$ and $H_{5\%}$ with theoretical cumulative frequencies of 1% and 5%. According to Table 2.3.3, $\dfrac{H_{1\%}}{\overline{H}}=2.421$, $\dfrac{H_{5\%}}{\overline{H}}=1.953$.

So
$$H_{1\%}=2.421\times 2.2=5.3(m)$$
$$H_{5\%}=1.953\times 2.2=4.3(m)$$

$H_{1\%}$ here is basically the same height as the larger wave height of the 100 continuous waves in Table 2.3.1 (its cumulative frequency equals 1%).

2.3.2 Wave return period calculation

(1) When there are long-term measured wave data (more than 15 years) at or near the site of the project, frequency analysis is generally carried out with the maximum annual wave height series of a certain cumulative frequency in the sub-direction to determine the design waves of different recurrence periods. When a wave in a certain direction needs to be analyzed statistically, the waves in the left and right directions (22.5°) of this direction should be taken as the waves in this direction for statistics.

When collecting data, the sampling method of annual maximum value is adopted to select the annual maximum value in a certain direction from the column of "wave height" of the observation record, which is represented by significant wave $H_{1/10}$. Therefore, the calculation of significant wave design is based on the fact that the historical wave measurement data of various coastal observation stations are basically obtained by shore optical

Chapter 2 Design Standards of Coastal Engineering Structures

wave measuring instruments. According to the specified observation method, the column of "wave height" is composed of 15 – 20 obvious large waves recorded in the time of 100 wave period, from which the largest 10 waves are taken out and equated. Therefore, the "wave height" in the observation record essentially means that its accuracy is higher than the "maximum wave height" column in the observation report.

(2) Arrange the collected wave height data in descending order. Table 2.3.4 shows the statistical examples of 23-year maximum wave height in direction E of an observation station.

Table 2.3.4 Wave high frequency calculation table

m	$H_{1/10}/\mathrm{m}$	$K_i = \dfrac{H_i}{\overline{H}}$	K_{i-1}	K_{i-1}^2	K_{i-1}^3	$P = \dfrac{m}{n+1} \times 100\%$
1	3.5	1.926	0.926	0.857	0.794	4.2
2	3.5	1.926	0.926	0.857	0.794	8.3
3	3.1	1.706	0.706	0.498	0.352	12.5
4	3.0	1.651	0.651	0.424	0.276	16.7
5	2.4	1.321	0.321	0.103	0.033	20.8
6	2.3	1.266	0.266	0.071	0.019	25.0
7	2.1	1.156	0.156	0.024	0.004	29.2
8	2.0	1.101	0.101	0.010	0.001	33.3
9	1.8	0.991	−0.009	0.000	0.000	37.5
10	1.6	0.881	−0.119	0.014	−0.002	41.7
11	1.6	0.881	−0.119	0.014	−0.002	45.8
12	1.6	0.881	−0.119	0.014	−0.002	50.0
13	1.5	0.826	−0.174	0.030	−0.005	54.2
14	1.5	0.826	−0.174	0.030	−0.005	58.3
15	1.4	0.771	−0.229	0.052	−0.012	62.5
16	1.3	0.715	−0.285	0.081	−0.023	66.7
17	1.3	0.715	−0.285	0.081	−0.023	70.8
18	1.3	0.715	−0.285	0.081	−0.023	75.0
19	1.3	0.715	−0.285	0.081	−0.023	79.2
20	1.1	0.605	−0.395	0.156	−0.062	83.3
21	1.1	0.605	−0.395	0.156	−0.062	87.5
22	0.9	0.495	−0.505	0.255	−0.129	91.7
23	0.6	0.330	−0.670	0.449	−0.301	95.8
Total	41.8			4.338	1.599	

The cumulative frequency P that is greater than or equal to a certain wave height can be calculated according to Eq. (2.2.7), where m is the sequence number of wave heights from large to small, and n is the total number of wave heights for statistics, as shown in column 7 of Table 2.3.4. The recurrence period T corresponding to the cumulative frequency P (%) is calculated by Eq. (2.2.1).

(3) The cumulative frequency P (%), which is greater than or equal to a certain wave height, is just the empirical frequency points. Due to the relatively short data age, only $T = \frac{100}{4.2} \approx 2.4$ a once wave height values can be obtained in the above example. If the wave height of 50 or longer reappearance period is to be deduced, it is necessary to use the theoretical cumulative frequency curve to extend. At present, Pearson Ⅲ curve is usually used as the theoretical curve for the calculation of design wave return period. The specific calculation method is described below.

1) Given n annual maximum wave height sequences H_i, the mean value $\overline{H}$ is:

$$H = \frac{1}{n} \sum_{i=1}^{n} H_i \tag{2.3.7}$$

In the wave height series shown in Table 2.3.4, $H_{1/10} = 1.8$ m.

2) The mean square error of n wave heights H_i can be calculated by the following formula when the data years are short:

$$\sigma = \sqrt{\frac{1}{n-1} \sum_{i=1}^{n} (H_i - \overline{H})^2} \tag{2.3.8}$$

σ reflects the dispersion degree of wave height as a random variable. The mean square error relative to the mean wave height is called the coefficient of deviation, and its formula is:

$$C_V = \frac{\sigma}{\overline{H}} = \sqrt{\frac{\sum_{i=1}^{n} (k_i - 1)^2}{n-1}} \tag{2.3.9}$$

In the formula, $k_i = \frac{H_i}{\overline{H}}$ is called modulus ratio coefficient.

The larger the C_V is, the greater the degree of relative change of the series; Otherwise, the data is relatively concentrated. In the example in Table 2.3.4, $C_V = \sqrt{\frac{4.438}{23-1}} \approx 0.449$.

3) In the wave height series, the distribution of the data around the mean value is asymmetrical. The skewness degree of data can be explained by the skewness coefficient C_S, which can be expressed as:

$$C_S = \frac{\sum_{i=1}^{n} (k_i - 1)^3}{(n-3) C_V^3} \tag{2.3.10}$$

$C_S = 0$ represents the symmetrical distribution of random variables in the average value, that is, normal distribution. The larger the value of C_S, the higher the degree of

Chapter 2 Design Standards of Coastal Engineering Structures

asymmetry of distribution, as in this example.

4) The cumulative frequency function or distribution function of Pearson Type III curve is:

$$P(x \geqslant x_i) = \int_{x_i}^{\infty} \left[\frac{\beta^\alpha}{\Gamma(\alpha)} (x - a_0)^{\alpha-1} e^{-\beta(x-a_0)} \right] dx \qquad (2.3.11)$$

Where, $\Gamma(\alpha) = \int_0^\infty x^{\alpha-1} e^{-x} dx$ is gamma function of α, $\alpha = \dfrac{4}{C_S^2}$, $a_0 = \overline{X}\left(1 - \dfrac{2C_V}{C_S}\right)$; $\overline{X}$ is the average value of the series.

For the convenience of use, Eq. (2.3.11) has been made into a special integral table. Only according to $\overline{H}$, C_V and C_S obtained above, the modular ratio coefficient k_P can be obtained from the integral table of Pearson E-type curve, where k_P represents the modular ratio coefficient corresponding to a certain accumulation rate $P(\%)$.

Then

$$H_P = k_P \overline{H}$$

Where, H_P is the wave height corresponding to the cumulative frequency $P(\%)$ to be deduced.

5) Using k_P and the corresponding $P(\%)$ value, the dots are drawn on the probability grid paper, and the connected smooth curve is the theoretical cumulative frequency curve under the condition of a certain C_V and C_S. Generally, C_V refers to the calculated value, and C_S is a certain multiple of S. See Fig. 2.3.2.

The accumulated experience frequency values in column 7 of Table 2.3.4 are also plotted in Fig. 2.3.2 and compared with the theoretical curve. It can be seen that when $C_V = 0.45$, $C_S = 2C_V$, the theoretical curve and the empirical frequency point are poorly fitted, so the left and right ends of the curve should be lifted up respectively. For this reason, the values of C_V and C_S both slightly increase, $C_V = 0.525$, C_S is still a multiple of C_V, $C_S = 4C_V$, then the redrawn curve and the empirical frequency points fit well, the average error is small, generally not adjusted.

Using the selected theoretical curve, the significant wave heights for a 50-year event or other recurrence period can be found in Fig. 2.3.2.

(4) The reason why Pearson Type III frequency curve is used to fit the empirical frequency points is that it is a frequency curve expressed by mathematical function, which does not mean that the annual extreme value distribution of wave height follows P-III distribution from the origin. P-III curve is still an empirical curve based on the distribution characteristics of the data. Therefore, when the P-III curve is used to fit the measured value is not ideal, other linear lines can be considered. At present, the commonly used linear lines are Poisson-Gumbel compound extreme value distribution, log-normal probability distribution, etc.

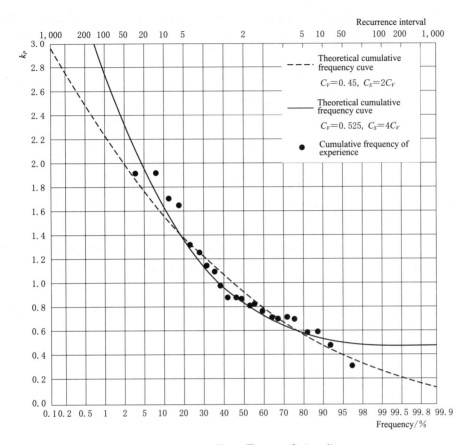

Fig. 2.3.2 Pearson Type Ⅲ curve fitting diagram

Exercise

Q1 Short answer: What is the basis for determining the design standard of seawall including tidal level and wave?

Q2 Fill in the blanks: The (   ) method is adopted to determine the design tidal level and design wave elements of coastal engineering. According to the relevant requirements of the *Code of Hydrology for Harbour and Waterway* (JTS 145—2015), at least (   ) extreme tidal level data and wave extreme value data are required, and the historical (   ) water level must be investigated.

Q3 Fill in the blanks: When the design tidal level is determined by annual frequency analysis method, the distribution of annual extreme tidal level is generally assumed to be (   ); In the estuary area affected by tides, the distribution of annual extreme tidal level is generally considered to be (   ); When the design wave is determined by annual frequency analysis method, the distribution of annual extreme anisotropic wave height is generally adopted to be (   ).

Chapter 2 Design Standards of Coastal Engineering Structures

Q4 Fill in the blanks: In general, annual frequency analysis is used to determine the designed tidal level of coastal engineering. If there is a shortage of data, but there are 5 years of tidal level data in the local area and more than 20 years of tidal level data in the adjacent waters, (   ) method can be used. For less than 5 years of tide data, but with more than 3 consecutive months of short-term tide (water) data including increased water, the (   ) method can be used; For important seawall projects, when measured tidal level observation data is lacking, temporary tidal level observation stations should be set up, and the observation period should not be less than (   ) years.

Q5 Short answer: Introduce the principle of "extreme synchronous difference method" to design high tidal level. What are the conditions of the two stations using this method?

Q6 Fill in the blanks: The return period is the tidal level that occurs once in 50 years. During the designed service life of the seawall for 50 years, the safety rate is (   ) and the risk rate is (   ).

Q7 Short answer: What is the design tidal level, the differences and similarities between the design tidal level of seawall engineering and the design tidal level of harbor engineering?

Q8 Short answer: The introduction to the estimation of wave recurrence period.

Chapter 3 The Interaction between Waves and Buildings

The interaction between waves and buildings, and coastal engineering design is closely related. Because of its importance, the research on this problem has been paid much attention. For example, in the design of breakwater, the wave force acting on the breakwater is the main external load, which is crucial to the safe operation of the breakwater and the cost of the project. In the practice of coastal engineering, there are many cases of building damage accidents due to insufficient understanding of wave characteristics. The worst disaster in 30 years to shock the world's coastal engineering community was the collapse of an outer dike in the Portuguese port of Sines, about 100 km south of Lisbon on Portugal's west coast. The new deepwater port will initially require the anchoring of 500,000 t of tankers. The breakwater will be on the west side of the port, extending 2 km south from the coast. This is the largest sloping breakwater ever built in the world. The outside of the breakwater is covered with 42 t of I-shaped blocks, and the maximum water depth is 50 m. Design wave height $H_s = 11$ m, wave period $T = 13.5$ s. In a storm on February 26, 1978, four parts of the breakwater were severely damaged, two of them (300 m and 150 m in length) had their superstructure completely collapsed, 1/3 of the I-shaped blocks disappeared from the surface of the water, and most of the I-block broke, and it is estimated that 2/3 of the I-block was lost. Restoration work began after the storm, but two storms occurred in December 1978 and February 1979, resulting in the loss of all the remaining and restored blocks. After the accident, several international investigations have been conducted, and many scholars and research institutions around the world have conducted special studies. This accident has aroused the attention of various countries to improve the design and building methods of sloping dikes, especially the method of determining the weight of blocks, the strength of blocks, and the action of irregular wave groups.

The interaction between waves and buildings usually includes three parts: the interaction between waves and sloping buildings, straight wall buildings (including steep wall buildings) and permeable buildings (pile column buildings). Diffraction phenomenon can also be regarded as a phenomenon of the interaction between waves and buildings.

Chapter 3 The Interaction between Waves and Buildings

3.1 The interaction between waves and sloping buildings

Sloping building is an old and traditionally widely used structure in coastal engineering. This kind of structure generally has low requirements on the foundation, which can use a large number of local materials. And when local damage occurs, it is easy to repair under the action of wind and waves, and generally does not produce destructive damage. The sloping embankment front rushing play is generally lighter than the vertical wall, and the wave impact force is smaller than the straight wall in the very shallow water area, so it is widely used.

The wave moves near the shore, but once it enters the sloping range, it will be deformed significantly due to the sudden reduction of water depth. Meanwhile, due to the influence of the friction of the surface of the slope and the backflow of the previous wave, the water body at the bottom advances slowly, making the front slope of the wave crest steeper, the back slope harder, the height of the wave crest higher, and finally breaking on the slope. The breaking wave water impinges on the slope, forming a larger impact force, and then the broken wave water climbs up the slope, producing a wave climb. The water climbs to a certain height and then waves fall back, and the cycle starts again. A curve that determines the reflected wave height is shown in Fig. 3.1.1.

There are two basic factors in wave and sloping interaction: wave breaking and reflection. When the slope is gentle, wave breaking on the slope plays a dominant role and reflection plays a secondary role. When the slope is steep (e.g., slope angle $\alpha \geqslant 45°$), wave reflection dominates, with little or no breaking. In engineering, it is customary to call the former case ($\alpha < 45°$) sloping type building, and the latter case is called steep wall or vertical ($\alpha = 90°$) building. Because wave reflection and breaking are mainly related to the wave sign besides the slope, the method of dividing upright and sloping buildings according to slope is sometimes not very exact from wave action characteristics.

In this section, the problems of wave reflection of sloping embankment, determination of stable weight of sloping embankment protection block, estimation of wave climbing height and excess wave amount, calculation of wave pressure of parapet wall, wave force of sloping, calculation of submerged embankment stability and wave elimination will be discussed. For the convenience of narration, the description of some problems is not strictly according to the distinction between sloping embankment and straight wall embankment.

3.1.1 Wave reflection

When a wave encounters a building in the process of propagation, it generates reflected waves. When the incoming wave acts on the building in a positive way, the reflected wave propagates in the opposite direction to the incoming wave. The wave superimposed

3.1 The interaction between waves and sloping buildings

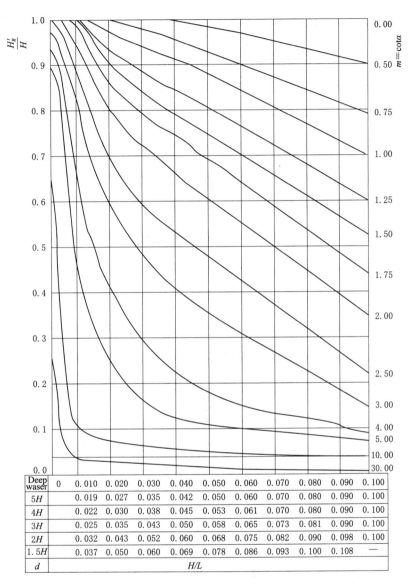

Fig. 3.1.1 A curve that determines the reflected wave height

by the reflected wave and the incident wave is called interference wave. Reflection degree of sloping surface is usually expressed by reflection coefficient K_R, which is defined as:

$$K_R = \frac{H_R}{H_I} \qquad (3.1.1)$$

Where, H_R is reflected wave height, m; H_I is incident wave height, m.

Miche, a French scholar, adopted La Grange Method in 1944 and theoretically obtained the reflection coefficient as follows:

$$K_R = K_\Delta \frac{L_0}{H_0} \sqrt{\frac{2\alpha}{\pi}} \frac{\sin^2\alpha}{\pi} \qquad (3.1.2)$$

47

Where, K_Δ is coefficients related to roughness and permeability of buildings, $K_\Delta < 1$, for smooth impervious wall, $K_\Delta = 1$; L_0 is deep water wavelength, m; H_0 is deep water wave height, m; α is slope angle, measured in radians.

Miche also obtained the limit wave steepness on a slope without breaking:

$$\delta_m = \left(\frac{L_0}{H_0}\right)_{max} = \sqrt{\frac{2\alpha}{\pi}} \frac{\sin^2\alpha}{\pi} \qquad (3.1.3)$$

When deep water wave steepness of incoming wave $\delta_0 = H_0/L_0$ is greater than δ_m, waves break, on the contrary, the waves do not break.

Iribarren and Nogales in Spain found that the critical slope of waves approaching breaking was:

$$\tan\alpha = \frac{8}{T}\sqrt{\frac{H}{2g}} \qquad (3.1.4)$$

The reflection coefficient $K_R = 0.5$ was verified by model experiment.

Miche's methods were used by standards in countries such as the former Soviet Union and Japan.

The reflection coefficient determination method suggested by *Code of Design for Breakwaters and Revetments* (JTS 154—2018) is based on Miche's theoretical solution, and a set of diagram for determining reflection coefficient K_R is drawn by collecting many laboratory test data at home and abroad, as shown in Fig. 3.1.2. In the figure, the y-coordinate is H'_R/H_1, H'_R is the reflected wave height of a smooth impervious wall ($K_\Delta = 1$), the steepness of the incident wave (deep water or local) can be determined from the known sloping ratio H'_R/H_1, and then the reflected wave height can be determined by the equation below:

$$H_R = K_\Delta H'_R \qquad (3.1.5)$$

Where, K_Δ is the roughness permeability coefficient related to the structure type of slope protection, determined according to Table 3.1.1.

Table 3.1.1   Roughness permeability coefficient K_Δ

Structural form	K_Δ
Whole piece smooth impervious veneer (asphalt concrete)	1.00
Concrete	0.90
Stone-laying	0.75 – 0.80
Block stone	0.60 – 0.65
Four-legged hollow block	0.55
Dimension stone	0.50 – 0.55
Concrete block	0.50
Four-legged cone	0.40
I-shaped block	0.38
Fence plate	0.49

3.1 The interaction between waves and sloping buildings

Seeling and Ahrens were recommended by the 1984 US Coastal Protection Manual.

The method was proposed in 1981 to determine the reflection coefficient, ξ (or Iribarren number) is defined as follows:

$$\xi = \frac{1}{\cot\alpha \sqrt{H_1/L_0}} \tag{3.1.6}$$

Where, α is slope angle; H_1 is incident wave height, m; L_0 is deep water wave length, m. After calculating ξ, K_R can be determined according to Fig. 3.1.2.

3.1.2 Stability calculation of sloping dike protection block

As mentioned above, the wave propagates to the slope, the water depth becomes shallower, and the wave deformation and reflection occur. When the embankment slope is steep, the wave generally only reflects and does not break; when the embankment slope is slow, the wave will break. When the broken wave flows up and down, the force will be generated on the protective block, resulting in the block losing stability. In general, there are three different types of instability that can occur in a veneer block (or block):

1) The block slides under the impact of water flow along the embankment surface.

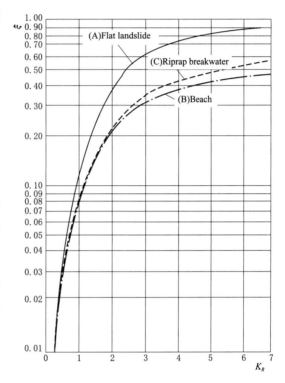

Fig. 3.1.2 $K_R - \xi$ curve (SPM)

2) When the wave falls back, the water in the dike seepage and the tidal level outside the dike drop quickly, resulting in the wave buoyant force, which makes the block extrude upward.

3) Surface blocks roll under the action of waves. There are many existing formulas for calculating block stability, and there are three mechanical schemata, corresponding to the above three instability modes. Of course, each author's assumptions and treatment methods are different, and the formulas obtained are also different. The following is a brief introduction.

(1) Iribarren formula. In 1938, Iribarren, a Spanish engineer, proposed a method to determine the stable weight of a block according to its sliding instability model. He believed that when the wave was broken on the slope, the jet impacted the slope vertically, and an upward force opposite to the direction of the jet was generated at the moment

Chapter 3 The Interaction between Waves and Buildings

the jet disappears. In this case, sliding instability may occur under the action of the self-weight component of the block (Fig. 3.1.2).

Let the bulk density of water and block (block stone) be r and r_b, A be the projected area of block stone in the vertical jet direction, and u be the jet velocity after wave breaking, which is regarded as the same as the water quality point velocity at the time of breaking. According to the theory of solitary wave, wave velocity $C_b = \sqrt{gd_b}$ at the time of breaking. Where d is the depth of crushing water, so we can obtain:

$$F = K_1 r d_b A \qquad (3.1.7)$$

The crushing water depth d_b is proportional to the crushing wave height H_b, then:

$$F = K_2 r A H \qquad (3.1.8)$$

Since A is proportional to $(W/r_b)^{2/3}$, where W is the weight of the block, then underwater mass $W' = K_v l^3 (r_b - r)$. According to the ultimate equilibrium condition $W' = F_q$, where the weight of the underwater block and the lifting force on the wave are equal, the following is obtained:

$$K_v l^3 (r_b - r) = C_q l^2 \frac{r}{K} H_b$$

Let $S = r_b / r$, the weight of the stone in the air $W = K_v l^3 r_b$, $l = (W K_v / r_S)^{1/3}$ into the above formula, then:

$$\frac{r_b^{1/3} H}{(S-1) W_b^{1/3}} = \frac{K(K_v)^{2/3}}{C_q} \qquad (3.1.9)$$

The right side of Eq. (3.1.9) is a comprehensive parameter, which is related to many factors, such as slope angle, wave elements (such as wave steepness H/L, relative water depth d/L), C_d, C_m (related to Reynolds number and absolute roughness Δ), block characteristics (such as r_b, porosity P, placing mode, roughness Δ, etc.) and other parameters. Several main parameters were selected by Hudson and expressed as:

$$\frac{r_b^{1/3} H}{(S-1) W_b^{1/3}} = f(a, H/L, d/L, D) = N_S \qquad (3.1.10)$$

Where, a is the slope angle and D is the instability rate of the block. When the riprap is more than one layer, the function should also include the influence of the number of layers. Hudson determined the form of function through laboratory tests and expressed N_S as:

$$N_S = a (\cot\alpha)^{1/3}$$
$$a = (K_D)^{1/3} \qquad (3.1.11)$$

Thus, the stable weight can be calculated by the formula:

$$W = \frac{r_b H^3}{K_D (S-1)^3 \cot\alpha} \qquad (3.1.12)$$

3.1 The interaction between waves and sloping buildings

Where, N_S is the stable number; K_D is the coefficient, which is related to the type of block, the mode of discarding, the number of layers and the allowable damage rate D. When $D=0$, it is no damage condition, and the corresponding wave height is $H_D = 0$. For the same block, K_D varies when the rate of instability changes, which means that if a certain degree of damage is allowed, K_D should be larger, that is, a smaller stable weight can be used. In 1982, Xue Hongchao expressed K_D as:

$$\frac{1}{K_D} = \frac{1}{K_{D=0}}(1+D)^{-0.462} \quad (3.1.13)$$

According to the experiment of Nanjing Hydraulic Research Institute and Hohai University, the power of Eq. (3.1.13) is -0.33 (for ripped-stone embankment).

Hudson formula is still widely used today. It is also used in many foreign norms and domestic port waterway hydrological norms of the Ministry of Communications. The K_D number and allowable instability rate adopted in port waterway hydrological norms are shown in Table 3.1.2.

Table 3.1.2 The allowable instability rate and K_D value of various blocks

Armour block	Configuration	D/%	K_D	Specification
Four-legged hollow block	One layer	0	14.0	
Block stone	One layer	0-1	5.5	
Four-legged cone	Two layers	0-1	8.5	
Twisted H block	Two layers	0	18.0	$H \geqslant 7.5$ m
		1	24.0	$H < 7.5$ m
Block stone	Two layers	1-2	4.0	
Square	Two layers	1-2	5.0	
Twist king block	One layer		24.0	

$$F = K_2 r H (W/r_b)^{2/3} \quad (3.1.14)$$

The underwater weight of the block is perpendicular to the slope and is parallel to the slope.

The underwater weight of the block perpendicular to the slope is $W'\cos\alpha$, and the underwater weight parallel to the slope is $W'\sin\alpha$.

According to the sliding equilibrium condition, we can obtain:

$$W'\sin\alpha = \mu(W'\cos\alpha - F) \quad (3.1.15)$$

$$W'\left(1-\frac{r}{r_b}\right)\sin\alpha = \mu\left[W\left(1-\frac{r}{r_b}\right)\cos\alpha - K_2 r H(W/r_b)^{2/3}\right] \quad (3.1.16)$$

Where, μ is the friction coefficient between blocks.

We can obtain:

$$W = \frac{K r_b H^3 \mu^3}{\left(\dfrac{r_b}{r}-1\right)^3 (\mu\cos\alpha - \sin\alpha)^3} \quad (3.1.17)$$

Chapter 3 The Interaction between Waves and Buildings

Iribarren took $\mu=1$ and got the following formula:

$$W=\frac{Kr_b H^3}{\left(\dfrac{r_b}{r}-1\right)^3 (\cos\alpha-\sin\alpha)^3} \quad (3.1.18)$$

Coefficient K is 0.015 for general block stone, and 0.023 when $d>0.06L$. The value of the other block is 0.019. Coefficient K is obtained according to the analysis results of two breakwater disaster cases.

Iribarren formula has been widely used, but there are many problems, mainly:

1) The coefficient K is proved by subsequent tests that K is not a constant and its value is related to $m=\cot\alpha$. For example, according to the test obtained by Hudson in 1952, when $m=1.5$, $K=0.008$, and when $m=2$, $K=0.017$.

2) The friction coefficient $\mu \neq 1$, μ is greater than 1 was verified by the Ehrlich formula.

3) Several hypotheses are worth discussing, such as whether the wave is completely broken, whether the jet is vertical slope, etc.

In 1965, in the general report of the 21st International Shipping Conference, the Iribarren formula was changed to the following form. The instability mode it describes is shown in Fig. 3.1.3.

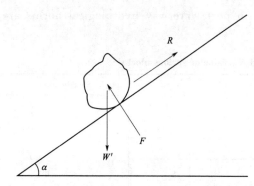

Fig. 3.1.3 Iribarren instability mode

$$W=\frac{Kr_b H^3}{\left(\dfrac{r_b}{r}-1\right)^3 (\mu\cos\alpha-\sin\alpha)^3} \quad (3.1.19)$$

In the formula, the empirical coefficient K is 0.43, 0.43 and 0.656 for blocks, blocks and four-legged cones, while μ are 2.38, 2.38 and 3.47, respectively.

To calculate the stable weight of the submerged protective block, Ehrlich used equivalent wave height instead of the design wave height in the formula. Equivalent wave height H' is calculated according to the following formula:

$$H'=\frac{\pi H^2}{Lc\,\text{th}\dfrac{2\pi z}{L}\text{sh}^2\dfrac{2\pi d}{L}} \quad (3.1.20)$$

Where, z is the distance between the location of the block and the water surface.

(2) Hudson formula. In 1959, according to the instability mode of upward ejection of block stone, American Hudson proposed another formula to calculate the stable weight of block stone, namely the famous Hudson formula. The basic assumption is that the wave is broken on the inclined plane, the water point velocity is equal to the wave velocity, and

3.1 The interaction between waves and sloping buildings

the velocity force acting on the block stone is calculated. The block loses stability due to the action of wave lifting force. At this time, the floating weight of the block in the water is equal to the upper holding force. Ignoring the friction between the blocks, the force acting on the block is as follows:

Velocity force (drag force):

$$F_d = \frac{1}{2} C_d K_a l^2 \frac{r}{g} u^2$$

Inertia force:

$$F_d = C_m K_v l^3 \frac{r}{g} \frac{\partial u}{\partial t}$$

Where, C_d and C_m are the velocity force and inertia force coefficients respectively; K_a and K_v are the area and volume coefficients respectively; l is the characteristic length of the block stone. The two forces can be obtained by combining them:

$$F_q = C_q l^2 \frac{r}{g} u^2$$

The particle velocity u, when the wave breaks, is equal to the wave velocity, so $u_b^2 = c^2 = g d_b = g H_b / K$, then:

$$F_q = C_q l^2 \frac{r}{K} H_b$$

The weight of the underwater block $W' = K_v l^3 (r_b - r)$.

According to the ultimate equilibrium condition $W' = F_q$ that the weight of the underwater block and the lifting force on the wave are equal:

$$K_v l^3 (r_b - r) = C_q l^2 \frac{r}{K} H_b$$

Let $S = r_b / r$, the weight of stone in the air is $W = K_v l^3 r_b$, substitute $l = [W/(r_b K_v)]^{1/3}$ into the above equation, then:

$$\frac{r_b^{1/3} H}{(S-1) W_b^{1/3}} = \frac{K (K_v)^{2/3}}{C_q} \tag{3.1.21}$$

The right side of Eq. (3.1.21) is a comprehensive parameter, which is related to many factors, such as slope angle, wave elements (such as wave steepness H/L, relative water depth d/L), C_d, C_m (related to Reynolds number and absolute roughness Δ), block characteristics (such as r_b, porosity P, placement mode, roughness Δ, etc.) and other parameters. Several main parameters were selected by Hudson and expressed as:

$$\frac{r_b^{1/3} H}{(S-1) W_b^{1/3}} = f(a, H/L, d/L, D) = N_s \tag{3.1.22}$$

Where, a is the slope angle and D is the instability rate of the block. When the riprap is more than one layer, the function should also include the influence of the number of layers. Hudson determined the form of the function through laboratory tests and expressed N_s as:

Chapter 3 The Interaction between Waves and Buildings

$$N_S = a(\cot\alpha)^{1/3}$$
$$a = (K_D)^{1/3} \tag{3.1.23}$$

Thus, the stable weight can be calculated by the formula:

$$W = \frac{r_b H^3}{K_D (S-1)^3 \cot\alpha} \tag{3.1.24}$$

Where, N_S is the stable number; K_D is the coefficient, which is related to the type of block, the mode of discarding, the number of layers and the allowable damage rate D. When $D=0$, it is no damage condition, and the corresponding wave height is $H_D=0$. For the same block, K_D varies when the rate of instability changes, which means that if a certain degree of damage is allowed, K_D should be larger, that is, a smaller stable weight can be used. In 1982, Xue Hongchao expressed K_D as:

$$\frac{1}{K_D} = \frac{1}{K_{D=0}} (1+D)^{-0.462} \tag{3.1.25}$$

According to the experiment of Nanjing Hydraulic Research Institute and Hohai University, the power of Eq. (3.1.18) is -0.33 (for ripped-stone embankment).

Hudson formula is still widely used today. It is also used in many foreign norms and domestic Ministry of Communications port and waterway hydrological norms. The K_D number and allowable instability rate adopted by port and waterway hydrological norms are shown in Table 3.1.2.

Although Hudson formula has been widely used, researchers from all over the world still put forward many critical opinions, mainly include:

1) There are many factors affecting the stable number N_S, and Hudson finally only boils down to two factors K_D and m. The coefficient K_D excludes the influence of many factors, but it is only related to the type of block, which is oversimplified. The most prominent one is ignoring the influence of wave period (or wavelength). Many experiments have proved that the influence of wave period exists. The influence of relative water depth is ignored.

2) The velocity force and inertia force are expressed as a function of u_2, which essentially ignores the effect of inertia force.

3) Other problems, such as the irregularity of waves (including the influence of group waves) and the influence of the permeability of the dike are not taken into account.

The effective wave height H_S is generally used in the design wave height of the application of Illibaron formula and Hudson formula. In recent years, many experiments have proved that H_S is unsafe, and it is more reasonable to adopt $H_{1/10}$ as the design wave. According to the hydrologic standard of port channel, effective wave height H_S is still adopted as the design wave height in general condition, and when $\frac{H}{D} < 0.3$, $H_{5\%}$ is

3.1 The interaction between waves and sloping buildings

adopted as the design wave height.

(3) Chaitsev method. Chaitsev followed the pattern of rolling instability in the block. As shown in Fig. 3.1.4, the formula for calculating stable weight of block stone is obtained. As shown in Fig. 3.1.5, the flow velocity when the water falls back is u, the characteristic length of the block is L, the resistance coefficient of the flow velocity is C_d, the area coefficient and volume coefficient of the block are K_a and K_v, then the velocity force acting on the block is $\frac{1}{2}C_d K_a l^2 \frac{r}{g} u^2$.

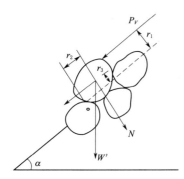
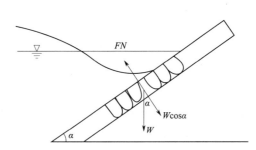

Fig. 3.1.4 Rolling instability mode   Fig. 3.1.5 Dry block stone instability diagram

The tangential (along the slope) and normal (perpendicular to the slope) components of the underwater weight of the block are $K_v l^3 (r_b - r)\sin\alpha$ and $K_v l^3 (r_b - r)\cos\alpha$.

According to the limit equilibrium condition of the rotation of the block stone around the fulcrum O:

$$\frac{1}{2}C_d K_a l^2 \frac{r}{g} u^2 r_1 = K_v l^3 (r_b - r)(r_2 \cos\alpha - r_3 \sin\alpha) \quad (3.1.26)$$

In the formula, r_1, r_2 and r_3 are moment arms (Fig. 3.1.5), and the stone weight $W = K_v l^3 r_b$ is substituted and sorted:

$$W = \frac{r_b}{\left(\frac{r_b}{r}-1\right)^3} \frac{u^6}{8g^3 K_v^2} \frac{(C_d K_a r_1)^3}{(r_2\cos\alpha - r_3\sin\alpha)} \quad (3.1.27)$$

Let the maximum flow rate u at the static water level be:

$$u = k_0 \sqrt{g} \sqrt[3]{H^2 L} \quad (3.1.28)$$

Then:

$$W = \frac{r_0 H^2 L}{\left(\frac{r_b}{r}-1\right)^3} \frac{k_0^6 (C_d K_a r_1)^3}{3 K_v^2 (r_2 \cos\alpha - r_3 \sin\alpha)} = \frac{r_0 H^2 L}{\left(\frac{r_b}{r}-1\right)^3} K_1 \quad (3.1.29)$$

The coefficient K_1 is obtained by the laboratory research of the Institute of Water Technology and Sanitation Engineering Science of the former Soviet Union:

$$K_1 = \frac{K}{\sqrt{1+\cot^3\alpha}} \quad (3.1.30)$$

Chapter 3 The Interaction between Waves and Buildings

Then:

$$W = \frac{Kr_b H^2 L}{\left(\dfrac{r_b}{r}-1\right)^3 \sqrt{1+\cot^3\alpha}} \tag{3.1.31}$$

For riprap, $K=0.025$. For throwing blocks, $K=0.017$.

Eq. (3.1.24) was adopted by Code CH 92—60. This formula reflects the effect of wavelength. Since the verification test of this formula was carried out when the wave steepness was large, it was later proved that when the wave steepness was large, Eq. (3.1.24) exaggerated the influence of wavelength. Subsequently, in the new specification published in 1986, Eq. (3.1.24) was revised as:

$$W = \frac{3.16 K_f r_b H^2}{\left(\dfrac{r_b}{r}-1\right)^3 \sqrt{1+\cot^3\alpha}} \sqrt{\frac{L}{H}} \tag{3.1.32}$$

In the formula, the coefficients K_f are set as 0.025 and 0.021 for the stone and concrete blocks of the throwing pile, 0.008 for the four-legged cone or other special-shaped blocks of the throwing pile, and 0.006 for the paved special-shaped blocks. The standard also specifies that the wave train accumulation rate of calculating the wave height is set as 2%.

The above introduces three representative methods for calculating the stable weight of the protective block of sloping dike, among which Hudson formula is more widely used. These formulas all appeared in the 1950s or earlier. Although they have been improved later, they are still obtained by regular wave mode.

Hudson formula and Iribarren formula did not consider the influence of wave period (or wavelength). Although the influence of wave steepness was considered, the test results could only be found within a certain range with a large wave steepness. Recent studies have proved that the influence of wave cycle change is related to embankment slope, i.e., broken-wave parameter ξ (also known as Iribarren number) can better reflect the combined influence of wave steepness and slope. Many tests have confirmed that the minimum stable number N_S value in the range of $\xi = 2 - 3$, indicating the worst block stability.

In recent years, some irregular wave tests have also shown that the stability tests of protected blocks under regular wave action, which used effective wave height as characteristic wave in the past, may get unsafe results. The regular wave height equivalent to the irregular wave is about 1.25 times of the effective wave height (this wave height is approximately equal to $H_{1/10}$). On the other hand, there are a lot of data on the influence of irregular wave groups, which show that the existence of wave groups reduces the stability of the block.

All the discussions mentioned before refer to the situation where waves act in a positive directions. What is the influence of oblique wave on the stability of the block? Recent studies

3.1 The interaction between waves and sloping buildings

have shown that oblique waves react differently to different shapes of blocks. Generally speaking, the influence of oblique wave on the stone shield is little, and for some special-shaped blocks (such as I-block), when the wave angle increases from 0° to 45°, the stability of the block decreases, and when the incident angle is greater than 45°, the stability increases again.

Finally, it is pointed out by the way that the units of W in Eq. (3.1.11), Eq. (3.1.16) and Eq. (3.1.24) are tons, and the units of r and r_b are t/m³. If the unit of bulk density is kN/m³, the right side of the formula should be divided by 9.81.

(4) Stability calculation of dry block stone protection. Dry block stone protection is also a common type of protection, its instability is generally due to the wave lifting force caused by the upward jump. The masonry thickness is generally determined in the calculation of dry block stone protection. The calculation method is derived from the balance of the normal forces (wave lifting force, gravity normal component, friction force, etc.) of the sloping surface.

As shown in Fig. 3.1.5, assuming that the normal buoyancy force acting on the sloping block is F_N, and the floating gravity of the block W'. If the thickness of the block is t and the cross-sectional area is S, the equilibrium condition of the normal force on the slope when the friction force is ignored is as follows:

$$F_N = W' \cos\alpha \qquad (3.1.33)$$

If the average pressure of the normal wave lifting force is p, then:

$$pS = tS(r_b - r)\cos\alpha \qquad (3.1.34)$$

In dimensionless form:

$$\frac{t}{H} = \frac{r}{r_b - r} \frac{p}{rH} \frac{\sqrt{1+m^2}}{m} \qquad (3.1.35)$$

Where, H is the calculating wave height, m.

If the pressure p acting on the block is known, the required masonry thickness t at ultimate equilibrium can be determined.

According to the above model, the hydrologic specification of port channel is determined through tests $[p/(rH)]$ value:

$$\frac{p}{rH} = 1.3(K_{md} + K_\delta) \qquad (3.1.36)$$

Then:

$$t = 1.3 \frac{r}{r_b - r} H(K_{md} + K_\delta) \frac{\sqrt{1+m^2}}{m} \qquad (3.1.37)$$

Where, K_{md} is coefficients related to sloping ratio $m = \cot\alpha$ and relative water depth d/H, listed in Table 3.1.3; K_δ is coefficients related to wavelength, listed in Table 3.1.4.

Chapter 3 The Interaction between Waves and Buildings

Table 3.1.3 K_{md}

d/H	K_{md}				
	1.5	2.0	3.0	4.0	5.0
1.5	0.311	0.238	0.130	0.080	0.054
2.0	0.258	0.180	0.087	0.048	0.031
2.5	0.242	0.164	0.075	0.041	0.026
3.0	0.235	0.156	0.070	0.037	0.023
3.5	0.229	0.151	0.067	0.035	0.021
4.0	0.226	0.147	0.065	0.034	0.020

Note: H in Eq. (3.1.37), when $d/L \geqslant 0.125$, $H=4\%$; when $d/L<0.125$, $H=13\%$.

Table 3.1.4 K_δ

L/H	10	15	20	25
K_δ	0.081	0.122	0.162	0.202

Eq. (3.1.37) is applicable to $1.5 \leqslant m \leqslant 5$.

Since the breaking wave parameters are closely related to the breaking and reflection of the wave in the sloping dike, it can be seen that it also has a certain relationship with the wave lifting force, which can be obtained from the analysis of the test data:

$$\frac{p}{rH} = 0.386 \xi^{1.88} \exp(-0.63\xi) \tag{3.1.38}$$

By substituting the above formula into Eq. (3.1.35), the masonry thickness t at limit equilibrium can be written:

$$t = 0.386 \xi^{1.88} \exp(0.63\xi) \frac{r}{r_b - r} H \frac{\sqrt{1+m^2}}{m} \tag{3.1.39}$$

$$\xi = \frac{1}{m\sqrt{H/L_0}} \tag{3.1.40}$$

The above formula is applicable for $\xi = 0.72 - 5.75$, $m = 1.25 - 5.0$, $L/H = 12 - 30$, $d/H = 1.4 - 5.2$. In terms of steep slope, the applicable scope is wider than Eq. (3.3.35).

Finally, it is pointed out that if the friction between the block and the lower block is considered when the block comes out, and the friction coefficient is set as μ, it is not difficult to derive:

$$\frac{t}{H} = \frac{r}{r_b - r} \frac{p}{rH} \frac{1}{\cos\alpha + \mu\sin\alpha} \tag{3.1.41}$$

Masonry thickness can also be determined as long as the size of the floating support is known. There are also formulas that take into account the friction between the blocks. Previous research. assumed the floating support force $p = K_1 rH$, and determined K_1 by visual observation of the stability of the block stone, and finally obtained by Eq. (3.1.41):

3.1 The interaction between waves and sloping buildings

$$t = 0.744 \frac{r}{r_b - r} H \frac{\sqrt{1+m^2}}{m+A} \left(0.476 + 0.157 \frac{d}{H}\right) \tag{3.1.42}$$

In Eq. (3.1.42), the coefficient A is 1.2 when the inclined joint is dry. When the flat seam is dry lined, 0.85 is taken. The prescribed Eq. (3.1.42) of harbor hydrology (Revised draft, 1994) applies $m = 0.6 - 1.5$, $L/H = 1.7 - 3.3$, $d/H = 12 - 25$.

3.1.3 Wave climb and fall depth

1. Conceptual framework

R_u is the vertical height of wave climbing along the slope calculated from the static tidal level, R_d refers to the vertical depth of wave falling calculated from the static tidal level and the wave climbing range R_{ud} refers to the sum of the climbing height and falling depth on the slope, namely, the variation range of wave surface before the inclined wall, as shown in Fig. 3.1.6.

The determination of sloping top elevation is closely related to wave climbing height. At present, the dike top elevation of dike engineering, bank protection and earth dam is determined according to the model of the design tidal level +

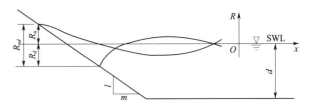

Fig. 3.1.6 The slope covered with waves

wave climbing height + safe elevation. Because of the square relationship between the height of the dike and the amount of engineering, so the size of the wave climb has a great influence on the amount of engineering.

Since the 1930s, many scholars have studied wave climbing. Until the 1960s, these studies were mainly on regular waves. After 1960s, the calculation methods of irregular wave climb began to appear in Soviet Code CH 288—64. At present, many calculation methods of irregular wave climb have been proposed. Regular wave research shows that the wave climbing height is related to the factors of the front wave, the characteristics of the embankment slope (slope, roughness and permeability), the water depth in front of the embankment and the characteristics of the beach.

The climb of natural wind waves on the dike slope is a random series of different sizes. Like the wave height, the climb can be studied by mathematical statistics, and the climb series can be characterized by statistical characteristic values. Similar to the statistical characteristic value of wave height, the expression method can define the cumulative frequency climb R_F, the mean value $R_{1/n}$ of partial large climb, the mean climb $\overline{R}$, etc.

In irregular climb height calculations, a statistical characteristic value for climb height can be related to a statistical characteristic value for wave height. After obtaining the climb height of a statistical characteristic value R_F, the climb height of other arbitrary

Chapter 3 The Interaction between Waves and Buildings

statistical characteristics can be converted depending on the statistical distribution of the climb height.

On the statistical distribution of climb, Battjes (1971) and Hong (1979) studied theoretically. According to the statistical analysis of measured data in Putian test station, Hohai University also proposed that Weibull distribution could be adopted for climb distribution, and the indoor irregular wave climb test conducted by Nanjing Hydraulic Research Institute also confirmed that irregular wave climb obews Weibull distribution. In foreign countries, such as SPM (1998 edition), the climb distribution adopts Rayleigh distribution.

2. Methods of hydrological specification for harbours and waterways

The hydrological specification methods of harbours and waterway are mainly based on regular wave test of indoor system. In this method, the calculation of wave climbing, wave falling amplitude is unified in a calculation model, and they are expressed as a function of slope number M (i. e., the Iribalen number Ir.). The method can be applied to the calculation of wave climb in the range of embankment slope variation. At present, the factors considered by this method are relatively complete in the calculation of regular wave climbing and falling.

Previous studies generally considered the influence of sloping coefficient $m = \cot\alpha$ and wave steepness H/L on the wave climbing under the action of forward regular waves, and determined the climb value according to these two factors. Later, many studies proved that the use of the Iribalen number can well reflect the characteristics of the interaction between waves and sloping dike, including wave climb, reflection, crushing and other problems. This method adopts the wave element in front of the dike and defines the number of slopes as:

$$M = \frac{1}{m}\left(\frac{L}{H}\right)^{1/2}\left(\tanh\frac{2\pi d}{L}\right)^{-1/2} \qquad (3.1.43)$$

When we have a period T or a deep water wavelength L, M can also be expressed as:

$$M = \frac{1}{m}\left(\frac{L_0}{H}\right)^{1/2} = \frac{T}{m}\left(\frac{g}{2\pi H}\right)^{1/2} \qquad (3.1.44)$$

R is used to represent the variation of tidal level of waves on the slope (positive from static tidal level upward), including climbing height R_u, falling depth R_d and climbing range $R_{ud} = R_u - R_d$, shown in Fig. 3.1.6. R can be shown in the following relation:

$$R = K_\Delta R_1 H \qquad (3.1.45)$$

Where, K_Δ is the roughness coefficient according to Table 3.1.1; H is the calculated wave height; $R_1(R_{u1}, R_{d1}, R_{ud1})$ is the value of wave climbing height, falling depth and climbing amplitude of the roughness coefficient $K_\Delta = 1$ and wave height $H = 1$ m. R_1 is related to the number of slopes M. Fig. 3.1.7 shows an example of the relationship between R_1 and M. R_1 can be determined by pressing the formula:

$$R_1 = K_1 \tanh(0.432M) + [(R_1)_m - K_2] R(M) \qquad (3.1.46)$$

3.1 The interaction between waves and sloping buildings

$$(R_1)_m = \frac{K_3}{2} \tanh \frac{2\pi d}{L} \left(1 + \frac{\frac{4\pi d}{L}}{\sh \frac{4\pi d}{L}}\right) \tag{3.1.47}$$

Where, the physical meaning of $(R_1)_m$ is the maximum climbing or falling depth corresponding to a relative depth d/L; K_1, K_2 and K_3 are the test coefficients obtained from Table 3.1.5.

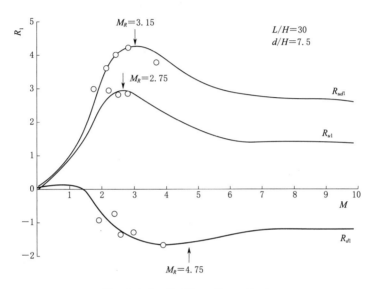

Fig. 3.1.7 $R_1 (R_{u1}, R_{d1}, R_{ud1})$

Table 3.1.5   Determine the wave climbing coefficient K_1, K_2, K_3

Coefficient	K_1	K_2	K_3
Wave climbing	1.240	1.029	4.980
Water depth	−0.760	−0.735	−2.760

When calculating climbing height and falling depth, $R_u(M)$ and $R_d(M)$ are used as $R(M)$ in Eq. (3.1.47), respectively:

$$(R_u)_M = 1.09 M^{3.32} \exp(-1.25M) \tag{3.1.48}$$

$$(R_d)_M = 0.350 M^{1.954} \exp(-0.420M) - 7.80 M^{2.02} \exp(-2.69M) \tag{3.1.49}$$

The climbing amplitude of wave on slope is calculated by R_{ud} according to the formula below $R_{ud} = R_u - R_d$.

The above regular wave method can be used to calculate the surge height in the case of no wind. Under natural conditions, it is often the action of wind-generated waves. Therefore, it is also necessary to consider the calculation of surge height under the direct action of wind. In order to apply the above calculation method of regular wave climb to the calculation of wind wave climb, the influence coefficient of wind speed should be introduced and the irregularity of wind wave climb should be considered. According to the comparison between the measured wind wave climb ($R_{u1\%}/H_{1\%}$) and the climb calculated according to the above regular wave method (R_u/H), the wind speed coefficient K_w can be determined. The measured data show that K_w is related to the relative wind speed W/C (C is the wave speed, W is the wind speed).

Under the direct action of wind, irregular wave climbing height can be calculated according to the following formula:

$$R_{u1\%} = K_\Delta K_w R_{u1} H_{1\%} \quad (3.1.50)$$

Where, $R_{u1\%}$ is wave climb of 1% cumulative frequency 1%, m; K_w is wind speed coefficient according to Table 3.1.6; R_{u1} is wave climb according to the regular wave method, the wave steepness adopted L/H_l.

Table 3.1.6　　　　　　　　　　Wind speed coefficient K_w

W/C	$\leqslant 1$	2	3	4	$\geqslant 5$
K_w	1.00	1.10	1.18	1.24	1.28

The cumulative frequency of climbing R_{u1} determined by Eq. (3.1.44) is 1%, and the climbing frequency of other accumulation rates $R_{u \cdot F}$ can multiply by the conversion coefficient of accumulation rate K_F:

$$R_{u \cdot F} = R_{u1\%} K_F \quad (3.1.51)$$

Table 3.1.7　　　　　　Conversion factor of cumulative climbing rate K_F

$F/\%$	0.1	1.0	2.0	4.0	5.0	10.0	13.7	20.0	30.0	50.0
K_F	1.17	1.00	0.93	0.87	0.84	0.75	0.71	0.65	0.58	0.47

Note: $F=4\%$ and $F=13.7\%$ is equal to $R_{u1\%}$ and $R_{u1/3}$.

The determination of wave climbing accumulation should be based on the characteristics and requirements of the building. It is generally believed that $R_{1\%} - R_{2\%}$ can be considered as basically no wave crossing.

The hydrological specification methods of ports and waterways is generally suitable for slope number $M=1-10$, $d/L=0.1-0.5$ and for vertical wall $m=\cot\alpha=0$, $M \to \infty$. At this time, $R_1 = R/H = K_1 = 1.24$ can be seen from Eq. (3.1.46), which is similar to the result calculated by vertical wave.

3. Putian Experimental Station method

The calculation method of storm wave climb in Putian Experimental Station is based on the statistical collation of measured wind wave climb data in Putian Seawall Test Station in Fujian Province. This method is adopted by the *Design Code for Rolled Earth-rock Fill Dams* (SDJ 218—84) of the Ministry of Water and Power, the national Standard *Code for Design of Levee Project* (GB 50286—2013), and the technical regulations of local sea reclamation projects in coastal provinces and cities.

Under the direct action of wind, the height of the normal wave on a single slope with

3.1 The interaction between waves and sloping buildings

$m = 1.5 - 5.0$ is calculated by the following formula:

$$R_F = \frac{K_\Delta K_v K_F}{\sqrt{1+m^2}}\sqrt{HL} \qquad (3.1.52)$$

Where, R_F is the wave climb value of cumulative rate F in the climb series, m; K is slope roughness permeability coefficient, as shown in Table 3.1.7; K_v is coefficient related to dimensionless wind speed $W/\sqrt{gd}$ (W is wind speed and d is water depth) determined according to Table 3.1.8; K_F is conversion coefficient of climbing accumulation rate, as determined by Table 3.1.9; H is average wave height of the waves in front of the dike, m; L is wave length in front of the dike, m.

Table 3.1.8　　　　　　　　　Coefficients K_v

$W/\sqrt{k_d}$	≤1.0	1.5	2.0	2.5	3.0	3.5	4.0	≥5.0
K_v	1.00	1.02	1.08	1.16	1.22	1.25	1.28	1.30

Table 3.1.9　　　　　　　　　Coefficients K_F

$\overline{H}/d$	$F/\%$	1	2	3	4	5	10	13	20	30	50
<0.1		2.23	2.07	1.97	1.90	1.84	1.64	1.54	1.39	1.22	0.96
0.1 - 0.3	K_F	2.08	1.94	1.86	1.80	1.75	1.57	1.48	1.36	1.21	0.97
>0.3		1.86	1.76	1.70	1.65	1.61	1.48	1.40	1.31	1.19	0.99

The observation range of the Putian Experimental Station method is as follows: $W/\sqrt{gd} = 1.39 - 6.65$, $H/d = 0.08 - 0.35$, $m = 1.5 - 3.0$.

In addition to the two methods discussed above, the most common is the USA Short Protection Manual (SPM) method and the former Soviet Code СНиП II 2.06.04 - 82 * method currently abroad. The SPM method is basically based on the diagram of $R/H - m$ obtained from the regular wave test in the room, for different water depths d/H'. Give different graphs. The former Soviet Code СНиП II 2.06.04 - 82 * is a calculation method that combines large number of regular wave test results and a small number of irregular wave test results (including using foreign data), and also gives the calculation method for regular wave without wind and wave climb respectively.

4. Calculation of wave climb of compound sloping dike section

The above is the case of a single slope, but in coastal protection engineering, the compound sloping dike section is widely used, the climb calculation of the compound section is discussed below.

The most common type of compound inclined dike is the one with platform. In general, the platform elevation is set near the design high tidal level for the purpose of reducing the height of wave climb. This is because the laboratory test confirmed that the platform located near the static tidal level of the wave elimination effect is the best. At the same time, the longer the width of the platform, the better the effect of wave reduction, but after reaching a certain length, the effect of further reducing the height of the platform is not obvious, so it is not economically worthwhile to make the platform too long.

(1) Reduced slope method. The basic starting point of the converted slope method is to seek the single slope degree m_e with the same climb value in the case of double slope through laboratory tests, and then determine the climb height according to the single slope method.

If the slope coefficient above and below the compound dike platform is m_u, m_d (Fig. 3.1.8), then:

$$\Delta m = m_d - m_u \tag{3.1.53}$$

1) When $\Delta m = 0$, that is, if the slopes above and below the platform are consistent, the slope m_e will be converted into the following formula:

$$m_e = m_u \left(1 - 4.0 \frac{|d_w|}{L}\right) K_b \tag{3.1.54}$$

Where, $|d_w|$ is the absolute value of the depth of the platform.

2) When $\Delta m > 0$, that is, the downhill is slow and the uphill is steep:

$$m_e = (m_u + 0.3\Delta m - 0.1\Delta m^2) \left(1 - 4.5 \frac{d_w}{L}\right) K_b \tag{3.1.55}$$

3) When $\Delta m < 0$, the downhill slope is steep and the uphill slope is slow:

$$m_e = (m_u + 0.5\Delta m + 0.03\Delta m^2) \left(1 + 3.0 \frac{d_w}{L}\right) K_b \tag{3.1.56}$$

In Eq. (3.1.25)- Eq. (3.1.27):

$$K_b = 1 + 3 \frac{B}{L} \tag{3.1.57}$$

Where, d_w is the water depth on the platform.

When the platform is below the static tidal level, the value is positive; otherwise, the value is negative. B is the platform width and L is the average wavelength (Fig. 3.1.8).

The converted slope method is obtained under the following test conditions: $m_u = 1.0 - 4.0$, $m_d = 1.5 - 3.0$, $d_W/L = -0.025 - 0.025$, $0.05 < B/L \leq 0.25$. In the preparation of seaway technical regulations, Zhejiang Provincial Water Resources Department has verified the converted slope meth-

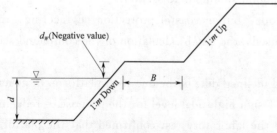

Fig. 3.1.8 Duplex slope with platform

od and believed that when m is $m_u \leqslant 0.4$ and $m_d < 1.5$, the climb value can also be calculated according to the converted slope m_e according to the test.

(2) T. Saville's method of imaginary slope. Sometimes complex cross sections, such as double platforms, are used in engineering design. At this time, the above-mentioned converted slope method cannot be calculated. The hypothetical slope method proposed by Sevier can be used in the case of complex cross sections.

The method replaces the compound slope with an imaginary single slope, and then determines the climbing height by trial calculation. The steps are as follows: the depth d_b of wave breaking is determined (Fig. 3.1.9). Take point B in the depth of the broken water as the starting point of the imaginary slope (point B may be on the slope or on the beach in front of the slope), arbitrarily

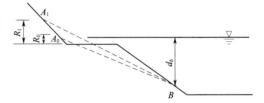

Fig. 3.1.9 T. Saville's method of imaginary slope

assume a climb value R_0, and its climb end point is A_0. Connect $A_0 B$ to obtain the imaginary slope $A_0 B$, and then determine the climb on the imaginary slope according to the single-slope method, which is R_1. In general, $R_1 \neq R_0$. Then, R_1 is taken as the imaginary climb value, and its climbing end point is A_1, and $A_1 B$ is connected to obtain a new imaginary slope. Then, the climb of section $A_1 B$ was determined by the single-slope method, and R_2 is obtained. Until the assumed climb is consistent with the calculated value, this climb is the desired value.

As for the calculation of the crushing depth d_b, Sevier originally suggested that the formula obtained by isolated wave theory should be followed:

$$d_b = 0.388 H'_0 \sqrt[3]{\frac{L_0}{H'_0}} \qquad (3.1.58)$$

Where, H_0 and L_0 are elements for calculating deep water waves.

Although the imaginary slope method can be used to compare the compound section in general condition, it still has some problems. It is verified that when the platform width is large, the results obtained by this method are much smaller, and the test value can reach 1.5 - 2.0 times of the calculated value. Then, when the foreshore is flat, the deviation of calculation results is also large. Therefore, some data indicate that this method is only applicable when $B/L < 0.15$ and the bottom slope is steeper than 1/30. At the same time, according to the calculation of Seville method, this method can well reflect the change of platform width and slope, but when the platform width and slope remain unchanged, the impact of platform elevation change on the climb can not be reflected, which is also inconsistent with the reality.

3.1.4 Estimation of overflow of sloping levees

As mentioned above, the height of the top of the seawall and the revetment is deter-

Chapter 3 The Interaction between Waves and Buildings

mined according to the design tidal level, the height of the wave climb and the safe elevation. However, even if the height of the dike is determined according to the design standard, there is the possibility of excessive waves under some abnormal meteorological conditions. On the other hand, from the economic perspective, it is not necessarily reasonable to design according to the standard of no more waves. If the design considers allowing a certain amount of waves, the engineering quantity of a certain section can be reduced, and the protection of the slope after the top of a certain embankment can be increased, and the safety of the embankment protection can be ensured, which may be more reasonable in the economy. Therefore, it is of great practical significance to calculate the amount of excess waves and the relationship between the amount of excess waves and the safety of embankments (namely to determine the so-called "allowable amount of excess waves").

At present, SPM method, Owen method, Battjes method and Hetian method are the most representative methods for calculating overpass quantity abroad. Among them, Hetian method is only applicable to the case of straight wall and is given in the form of many curves. Owen method, which is relatively simple to calculate, is introduced as follows.

M. W. Owen of the UK Hydraulic Test Station carried out a large number of tests on single slope and multiple slope sections under the condition of irregular waves and obtained a set of empirical charts. The test range is slope $m=1-4$, embankment top height $h_1 = 1-3.0$ m above static tidal level, water depth on compound platform $d_W = 0-4$ m, platform width $B = 0-8$ m. The effective wave height $H_S = 0.75 - 4.0$ m, the wave steep $H_S/L_0 = 0.035 - 0.055$, the wave direction Angle $\beta = 0°$ (forward wave)$- 60°$, the test spectrum is JONSWAP spectrum, the seabed slope in front of the embankment is $1:20$, the water depth in front of the embankment is $d=4$ m, the model scale is $1:25$.

If the dimensionless surge quantity $Q^* = \dfrac{Q}{TgH_S}$ is taken, Q is the average surge quantity of a single width [m³/(s·m)], T is the average period of the trans-zero wave, and the height of the top of the dimensionless dike $h_c^* = \dfrac{h_c}{T\sqrt{gH_S}}$ is taken, and h_c is the height of the top of the dike above the static tidal level, then the following relation exists in both single slope and multi-slope sections:

$$Q^* = A\exp(-Bh_c^*) \quad (3.1.59)$$

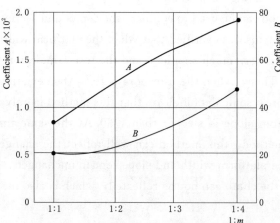

Fig. 3.1.10 Coefficients A, B

3.1 The interaction between waves and sloping buildings

Where, A and B are empirical coefficients. Analysis shows that the above equation has a good correlation (correlation coefficient 0.97). The values of A and B in the case of single slope are shown in Fig. 3.1.10. The above is the case of forward wave action. In case of oblique wave action, A and B should multiply by the correction coefficient in Fig. 3.1.12. Fig. 3.1.11 shows that when the wave direction deviates from the straight-on direction by approximately 15°, the overtopping volume increases.

All the cases discussed above belong to the optical landslide surface. For the roughness slope, the coefficient B in Eq. (3.1.59) should be calculated as (B/K_Δ), and K_Δ is the roughness coefficient.

There is little information available on the study of the allowable wave overtopping because it is difficult to determine the allowable wave overtopping under laboratory conditions. According to the example of seawall damage, Japan's Ata estimated the allowable surge as shown in Table 3.1.10, which was included in the *Design Standard for Port Buildings* compiled by the Japan Harbor Association. In addition, according to the survey conducted by Fukuda, Japan, the relationship between the amount of overtopping waves and the safety of pedestrians, cars and houses is shown in Table 3.1.11. The figure in the table refers to the value of the amount of overtopping waves immediately after the dike. If it is 10 m away from the dike, the value can enter a single digit.

In the case of wave overtopping, appropriate drainage measures should be considered to ensure the safety of levees and bank protection.

Table 3.1.10　　　　　　　　　　Allowable wave overtopping

Type	Surface protection	Overtopping/[m³/(s·m)]
Dike	No protection on the top and inner slope of the levee	Below 0.005
	The top of the levee is protected, but the inner slope is not protected	0.02
	There are protection on three sides	0.05
Revetment	The top of the levee is not protected	0.05
	The top of the levee is protected	0.20

Table 3.1.11　　　　　　　　　　Safety behind the dike

Object	Dike distance	Overtopping/[m³/(s·m)]
Pedestrian	Right behind the dike (50% safety)	2×10^{-4}
	Right behind the dike (90% safety)	3×10^{-5}
Automobile	Right behind the dike (50% safety)	2×10^{-5}
	Right behind the dike (90% safety)	3×10^{-6}
House	Right behind the dike (50% safety)	7×10^{-5}
	Right behind the dike (90% safety)	1×10^{-6}

3.1.5 Wave force on a slope

The wave pressure acting on a slope discussed in this section refers to the case where the slope protection is monolithic or prefabricated concrete or reinforced concrete slab. The design of slope protection of concrete slab includes two parts: The calculation of the strength of the protection panel and the calculation of the overall stability of the protection panel. The wave load of the former is the impact pressure when the wave strikes the slope protection, while the latter is the buoyancy force of the wave (generally occurring when the wave falls back).

1. The impact pressure of the wave

The impact pressure of the wave on slope includes the position and value of the maximum impact pressure, the distribution of pressure along the slope, the statistical characteristics of the impact pressure and so on.

The former Soviet Union Zhongkovsky as early as 50 years according to the principle of free jet flow is obtained when the waves hit the slopes of hit point location and maximum pressure. The basic assumptions are as follows: When the wave is broken, the water body at the summit is shot horizontally by free jet, and shoots to the slope under the action of gravity.

The initial velocity of the horizontal jet V_A is equal to the sum of the horizontal component V_x of the trajectory velocity of the fluctuating water quality points and some wave velocities, namely $V_A = V_x + nc$ (n is the coefficient, c is the wave velocity). Under these assumptions and with the introduction of some empirical coefficients, the maximum wave impact pressure p_B and the horizontal coordinate of the impact point X_B obtained by Zhong's conduction are:

$$p_B = 1.7 \frac{r}{2g} V_B^2 \cos^2 \varphi \tag{3.1.60}$$

$$X_B = \frac{1}{g} \left[-\frac{V_A^2}{m} + V_A \sqrt{\left(\frac{V_A}{m}\right)^2 + 2gy_0} \right] \tag{3.1.61}$$

See Fig. 3.1.11 (a) for other symbols, where y_0 is composed of the breaking water depth d_{cr} of the wave on the slope and the wave surface height η at the time of the wave breaking. According to Hashachev's test formula:

$$d_{cr} = H \left(0.47 + 0.023 \frac{L}{H} \right) \frac{1+m^2}{m^2} \tag{3.1.62}$$

$$\eta = H \left[0.95 - (0.84m - 0.25) \frac{H}{L} \right] \tag{3.1.63}$$

Where, H and L are wave height and wave length, $m = \cot\alpha$.

In the early 1970s, based on the results of large-scale flume tests in the ВОДГЕО of Soviet Union, the Soviet Code CHu Ⅲ 57-75 and СНиП Ⅱ 2.06.04-82 * proposed the following new formula for calculating wave shock pressure:

$$p_B = K_1 K_2 \overline{p} r H \tag{3.1.64}$$

$$K_1 = 0.85 + 4.8 \frac{H}{L} + m \left(0.028 - 1.15 \frac{H}{L} \right) \tag{3.1.65}$$

3.1 The interaction between waves and sloping buildings

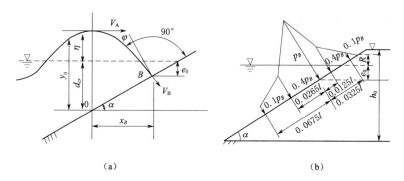

Fig. 3.1.11 The wave pressure of a wave hitting a slope

The coefficient K_2 is related to the wave steepness L/H, determined by Table 3.1.12, $\bar{p}$ is the relative maximum wave pressure, and is related to the absolute value of tidal height, determined by Table 3.1.13.

Table 3.1.12   Coefficient K_z

L/H	10	15	20	25	35
K_Z	1.00	1.15	1.30	1.35	1.48

Table 3.1.13   Relative maximum pressure $\bar{p}$

Wave height H/m	0.5	1.0	1.5	2.0	2.5	3.0	3.5	$\geqslant 4.0$
$\bar{p}$	3.70	2.80	2.30	2.10	1.00	1.80	1.75	1.70

The depth e_0, where the maximum wave pressure operating point B is below the static water level, is calculated according to the formula below:

$$e_0 = d_{cr} + \frac{1}{m^2}(1 - \sqrt{2m^2 + 1})(d_{cr} + \eta) \quad (3.1.66)$$

In the formula, d_{cr} and η are determined by Eq. (3.1.62) and Eq. (3.1.63) respectively. It is not difficult to verify that Eq. (3.1.66) is derived based on the principle of free jet flow and the initial jet velocity $V_A = \sqrt{g(\eta + d_{cr})}$. However, it should be pointed out that the value of e_0 calculated according to Eq. (3.1.66) may be negative in the case of steep slope, that is, the striking point is above the static tidal level, which is inconsistent with the test data obtained so far. Therefore, when the value of e_0 is negative, it is more reasonable to take $e_0 = 0$.

The pressure distribution along the slope is shown in Fig. 3.1.11 (b). In the figure, R is the climbing height and l is the calculation length, which is calculated according to the following formula:

$$l = \frac{mL}{\sqrt[4]{m^2 - 1}} \quad (3.1.67)$$

The above method is obtained based on regular wave test. When used to calculate the strength of concrete slabs, $H_{1\%}$ is generally taken as the calculated wave height. At present, there are a few calculation methods for irregular wave patterns, such as the method obtained by ШАЙТАН from the former Soviet Union based on the field observation data of reservoir, but which have not yet been incorporated into the standard. The rule wave method is also adopted in the standard of earth-rock dam.

2. Wave lifting force

In the process of fluctuation, the wave pressure acting on the panel upward is called buoyancy support. In fact, at any given moment, the wave pressure is alternating between positive and negative over a sufficient length of the slope. However, in a small section studied, the resultant wave pressure on both sides of the panel may be upward or downward for a certain instant. The test shows that the maximum upward wave pressure usually occurs in a certain section of the inclined plane near the lowest position (the moment before the next wave strikes the dike) when the wave surface falls back to the lowest position (the bottom) of the wave surface curve in front of the dike.

The wave lifting force is related to the wave element in front of the embankment, the relative change of tidal level inside and outside the embankment, the slope degree of the embankment, the permeability of the protective surface and cushion. Different from the wave striking force, the relative change of tidal level inside and outside the dike has a great influence on the wave lifting force. At the same time of wave climbing, part of the water flows into the cushion through the gap, and the wave begins to fall back after climbing to the highest point. In the process of falling back, the water in the cushion goes down along the cushion, or drains outward through the gap in the protective surface. It is confirmed by observation that the dynamic tidal level in the cushion lags behind the fluctuation tidal level outside the embankment, that is, the tidal level in the cushion is slightly lower than the tidal level outside the embankment when the wave climbs. On the contrary, when the wave falls back, the tidal level in the dike is slightly higher than the tidal level outside the dike due to the small permeability velocity, and the tidal level difference is also an important reason for the formation of wave buoyancy.

The former Soviet Code СНиП Ⅱ 2.06.04 − 82 * recommended wave lifting force are calculated as follows:

$$p_f = K_1 K_2 \overline{p}_u r H \tag{3.1.68}$$

Where, K_1 is according to Eq. (3.1.65); K_2 is according to Table 3.1.12; $\overline{p}_u$ is the relative wave lifting force determined according to Fig. 3.1.12.

Eq. (3.1.68) is obtained under the condition of slit plate and permeability of 1.5%. As can be seen from the figure, the pressure diagram of floating support is a stepped form. It should be noted that the figure is the envelope diagram of the maximum floating support. In the figure, B is the length of the panel along the slope, and L is the wavelength.

3.1 The interaction between waves and sloping buildings

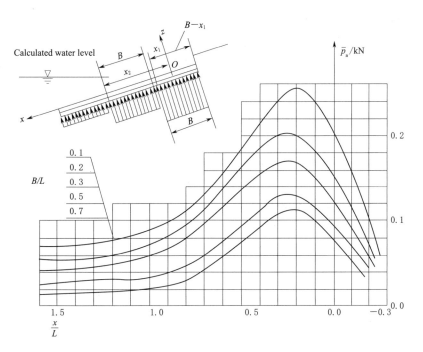

Fig. 3.1.12 Determine the inverse relative wave pressure curve

The thickness of the panel required for overall stability can be determined with fluctuating wave lifting force.

When determining the thickness of the panel, wave elements, slope ratio m and panel length B are used directly to calculate the panel thickness instead of the above floating support force. For example, the Soviet Code CH 288—64 uses the following formula to determine the panel thickness t:

$$t = 0.07 K H \frac{r}{r_b - r} \frac{\sqrt{1+m^2}}{m} \sqrt{\frac{L}{B}} \tag{3.1.69}$$

Where, K is coefficient, $K=1$ for monolithic block guard panel, $K=1.1$ for prefabricated guard panel; H is dike front wave height, m, generally acceptable $H_{1\%}$; L is wave length, m; B is the length of the protection panel along the slope direction (perpendicular to the water boundary line), m; r_b, r are panel and water bulk density, kN/m³; m is slope ratio refering to slope angle, (°).

3. Wave force at the top parapet of a sloping embankment

The section form with parapet at the top of sloping dike is widely used in breakwater and seawall. The wave force acting on the parapet of sloping dike is a very complicated problem. Besides the wave factor, there are also the scale factor (shoulder width b and slope ratio m) and roughening characteristics of sloping dike (or high foundation bed), water depth factor (water depth d_1 and water depth d in front of the dike), and the slope of beach in front of the dike. Under certain wave elements and sloping dike scale, water

Chapter 3 The Interaction between Waves and Buildings

depth change in front of wall plays a key role in wave breakage and wave force. Generally speaking, when the bottom of the parapet wall is higher than the static water surface, the wave force is small. With the rise of the static tidal level, the wave force increases rapidly. When the relative water depth in front of the wall is 0.9 – 5, the wave force reaches its maximum value, and the water depth in front of the wall continues to increase, the total wave force begins to decline and finally tends to a fixed value. The d_1/H value of the maximum total wave force is related to the wave arrival condition and the scale factor of the slope embankment. In the past, the calculation of wave force on the parapet wall of slope dike generally only considered three factors: wave steepness, water depth in front of the wall and water depth in front of the dike. Recent studies show that the influence of the width of foundation shoulder and slope ratio m cannot be ignored.

(1) Non-embedding of the parapet. The research results of Hohai University are introduced below. In this method, the total wave force on the parapet is expressed as:

$$\frac{P}{rH^2}=f\left(\frac{Hd_1}{L}\frac{d}{d}\frac{b}{H}\frac{}{H}m\right) \tag{3.1.70}$$

The analysis shows that the variation law of total wave force is very complicated when single factor is adopted. Through a large number of trial calculations, the complex factor S can be found to represent the total wave force, and better results can be obtained. The complex factor S is:

$$S=\left[\frac{d_1}{d}\left(\frac{d}{H}\right)^{2\pi H/L}+0.625\right]\left(\frac{L}{H}\right)^{0.1723} \tag{3.1.71}$$

In deep water, $d_1/d=1$, $d=0.5L$, then:

$$S_0=\left[\left(\frac{1}{2}\frac{L}{H}\right)^{2\pi H/L}+0.625\right]\left(\frac{L}{H}\right)^{0.1723} \tag{3.1.72}$$

When the total horizontal force on the wall peaks, $S=S_m=1.456$.

Since the height of the parapet of the sloping dike is not large, the wave pressure can be approximately considered to be evenly distributed, and the average pressure $\bar{p}$ can be calculated according to the measured total horizontal wave force and pressure acting height.

Under the condition that there is no artificial block cover in front of the wall and the bottom of the wall is not buried in the base bed, the average wave pressure acting $\bar{p}$ on the parapet is:

$$\bar{p}=(\bar{p}_1+\bar{p}_2)K_dK_bK_m \tag{3.1.73}$$

$$\bar{p}_1=1.017\left(\frac{1}{2}\frac{L}{H}\right)^{-4.10\frac{\pi H}{L}}\text{th}\left(\pi\frac{S}{S_0}\right)rH \tag{3.1.74}$$

$$\bar{p}_2=\begin{cases}1.402\left(\frac{H}{L}\right)^{0.792}\left(\frac{S}{S_m}\right)^{-b_1}\left[\frac{S}{S_m}\exp\left(1-\frac{S}{S_m}\right)\right]^{b_2}rH & S\geqslant 1 \\ 1.402\left(\frac{H}{L}\right)^{0.792}\frac{0.939^{b_2}}{0.687^{b_1}}SrH & S<1\end{cases} \tag{3.1.75}$$

3.1 The interaction between waves and sloping buildings

$$b_1 = 2.36 - 9.31\left(\frac{H}{L}\right) \quad (3.1.76)$$

$$b_2 = 7.17\exp\left(0.0555\frac{H}{L}\right) \quad (3.1.77)$$

Where, $\bar{p}_1$ and $\bar{p}_2$ are the reflection component and the influence component of base bed of average wave pressure, kN/m².

When the static tidal level is lower than the bottom of the parapet (Fig. 3.1.13), d_1 is negative; H and L are design wave height and wavelength, m; K_d, K_b and K_m are the influence coefficients of water depth in front of the dike, width of foundation shoulder and slope ratio, which are determined according to the following formula or Fig. 3.1.14 and Fig. 3.1.15.

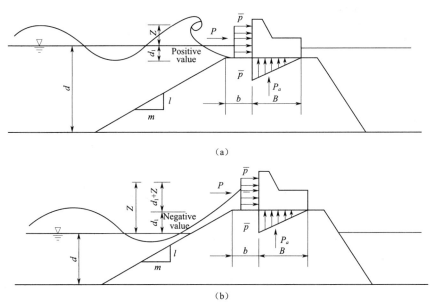

Fig. 3.1.13 Wave pressure profile of the parapet

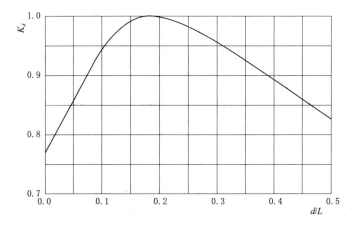

Fig. 3.1.14 $K_d - d/L$

Chapter 3 The Interaction between Waves and Buildings

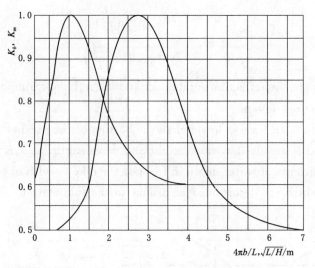

Fig. 3.1.15 $K_b - 4\pi \dfrac{b}{L}$, $K_m - \dfrac{1}{m}\left(\dfrac{L}{H}\right)^{1/2}$

$$K_d = 18.18\left(\dfrac{d}{L}\right)^{1.63} \exp\left(-8.53\dfrac{d}{L}\right) + 0.76 \qquad (3.1.78)$$

$$K_b = 5.78\left(4\pi\dfrac{b}{L}\right)^{2.67} \exp\left[-2.67\left(4\pi\dfrac{b}{L}\right)\right] + 0.60 \qquad (3.1.79)$$

$$K_d = 0.453\left[\dfrac{1}{m}\left(\dfrac{L}{H}\right)^{0.5}\right]^{8.85} \exp\left[-3.12\dfrac{1}{m}\left(\dfrac{L}{H}\right)^{0.5}\right] + 0.50 \qquad (3.1.80)$$

Where, b is the width of the base shoulder; m is the slope ratio.

The height of wave pressure on the parapet (d_1+Z) is determined by pressing the formula below:

$$(d_1+Z) = \left(1.235\dfrac{L}{H} - 1.595\right)\left(\dfrac{S}{1.456}\right)^{b_1} H \qquad (3.1.81)$$

The above formula applies to $S \leqslant S_B$, and S_B is calculated according to the following formula:

$$S_B = \left[2.76\left(\dfrac{H}{L}\right)^{0.701} + 0.625\right]\left(\dfrac{L}{H}\right)^{0.1723} \qquad (3.1.82)$$

The total horizontal wave force $P(\text{kN/m})$ on the parapet per unit length is:

$$P = \overline{p}(d_1+Z) \qquad (3.1.83)$$

The total wave lifting force $P_u(\text{kN/m})$ on the parapet per unit length is:

$$P_u = \mu\dfrac{B\overline{p}}{2} \qquad (3.1.84)$$

Where, saturation coefficient $\mu = 0.9$.

(2) The parapet wall is embedded. When the parapet wall is embedded in the base bed, it is generally believed that the wave pressure of the embedded part is reduced due to the shielding effect of the base bed. However, there are few research data on this issue. A

3.1 The interaction between waves and sloping buildings

simplified treatment method is proposed by referring to some similar regulations abroad and a few domestic test results in the revised draft of port channel hydrological standard.

As shown in Fig. 3.1.16, when there is no shielding block in front of the wall and the depth of the parapet embedded into the foundation bed is $h_b = (0.25 - 0.50)H$ and $h_b \leqslant 0.3 h_w$, the average pressure and the height of the pressure acting on the slope shoulder are still calculated according to the condition of no embedment, and the pressure of the embedded part of the parapet decreases linearly until the pressure at the bottom of the parapet is $P_b = 0.7\bar{p}$, The total horizontal force and the total wave lifting force were cal-

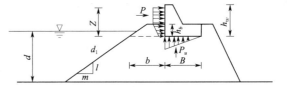

Fig. 3.1.16 Wave pressure distribution diagram when the parapet is embedded in the base bed

culated according to the pressure distribution diagram, and the saturation coefficient was $\mu = 0.9$.

(3) An artificial slope in front of the parapet. As shown in Fig. 3.1.17, when artificial block protection is used for sloping dikes, the protection usually extends to the parapet at the top of the dike, providing some protection for the parapet. The arrangement of the block in front of the parapet has two kinds: full cover (the slope top of the block protection is not lower than the top of the parapet) and half cover (the slope top of the block is lower than the top of the parapet). The half cover arrangement will increase the wave pressure in some tidal level conditions, so it should be careful.

For the wave pressure calculation of the parapet wall under full cover when two layers, two rows and four legs pyramid-shaped or twisted I-block are placed within the range of the front slope shoulder of the wall as shown in Fig. 3.1.17, the wave force can be calculated according to the condition without cover first (base shoulder width is calculated according to the dam without cover, as shown in Fig. 3.1.17), and then multiplied by the horizontal force reduction coefficient K_H and vertical force reduction coefficient K_V.

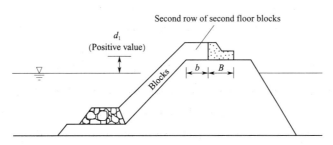

Fig. 3.1.17 A parapet covered by a block

For tetragonal vertebrae:

$$K_H = 0.2 d_1 / H + 0.43 \quad (3.1.85)$$

75

$$K_V = 0.2d/H + 0.61 \tag{3.1.86}$$

For the twisted I-block:

$$K_H = 0.62d_1/H + 0.53 \tag{3.1.87}$$

$$K_V = 0.62d_1/H + 0.77 \tag{3.1.88}$$

The above reduction coefficient of the wave pressure of the parapet with shielding block is proposed based on the experimental results of Hohai University, and the test conditions are as follows: $d_1/H = -0.5 - 0.5$, $m = 1.5$, $B = 1.0H$, $L/H = 15 - 25$. In all the tests with $d_1/H < 0.25$, K_H and K_V are less than 1.0. In the torsion font test, $K_H > 1.0$ once appeared when $d_1/H = 0.5$. According to this situation, the standard limits the use of the above method within $d_1/H \leqslant 0.3$, when $d_1/H > 0.3$, it should be determined by the test.

3.1.6 Wave attenuation characteristics and stability of submerged dikes

The sloping dikes discussed above actually refer to those that are out of water and generally do not wave overtopping. In this section submerged dikes whose tops are often below the static tidal level are discussed. The characteristics of wave action on water levee and submerged levee are different. When the water comes out of the dike, the wave usually breaks in front of the dike and impinges the slope with jet. In the case of submerged levee, when the water depth at the top of the levee is relatively small, the wave breaks when it crosses the levee. The form of breaking gradually transite from the roll type to the collapse type when the water depth at the top of the levee gradually increases, until finally no breaking occurs. At this time, the whole wave cross the levee and the wave impact basically disappear.

At present, the research on the effect of waves on submerged levees mainly includes the characteristics of wave attenuation of submerged dikes, that is, the height and wavelength of the transmitted wave behind the levees. On the other hand, the stability and force characteristics of submerged dikes under the action of waves are studied.

1. Wave attenuation characteristics of submerged levees

Wave elements, geometric scale of submerged levees, water depth and so on affect the characteristics of submerged levees. Among them, the relative position of submerged dike top elevation and static tidal level are the most important.

As shown in Fig. 3.1.18, a is the height from the static tide level to the top of the levee. When the water is discharged, a is a positive value. According to the ripped-submerged dike test results of Nanjing Hydraulic Research Institute and Hohai University, when $a/H \geqslant 1.0$, waves break on the front slope of the dike, and basically do not break the waves. The waves behind the dike are mainly formed by waves penetrating through the void, and the wave height behind the dike is very small. When $a/H = 0.25 - 0.5$, the water body after breaking can cross the dike into the rear. When $a/H = -(0.25 - 0.5)$, the overpass becomes the main component of the wave behind the dike, and the wave reflec-

3.1 The interaction between waves and sloping buildings

tion in front of the dike decreases obviously. When $a/H < -0.5$, the wave breakage is not obvious, and there is spray on the top, and the wave damping effect decreases. When $a/H < -1.5$, waves over the dike, there is basically no breakage, and there is little difference between the height of waves behind the dike and the height of waves in front of the dike. It is also found in the test that when $a/H = -1.5$ to -1.2, the height of waves behind the dike increases slightly.

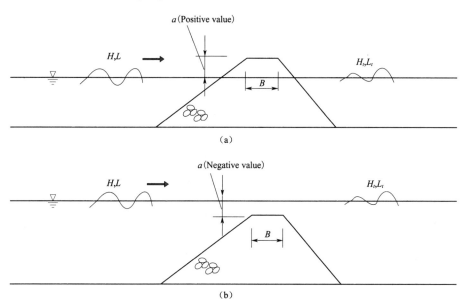

Fig. 3.1.18 Back wave element of off-shore dike

According to the test results of Nanjing Hydraulic Research Institute and Hohai University, the formula for calculating the height of wave height after riprap dike is as follows:

Submerged ($a/H \leqslant 0$):

$$\frac{H_t}{H} = \text{th}\left[0.8\left(\left|\frac{a}{H}\right| + 0.038\frac{L}{H}K_B\right)\right] \quad (3.1.89)$$

Outlet dike ($0.25 > a/H > 0$):

$$\frac{H_t}{H} = \text{th}\left(0.030\frac{L}{H}K_B\right) - \text{th}\left(\frac{a}{2H}\right) \quad (3.1.90)$$

$$K_B = 1.5\exp(-0.4B/H) \quad (3.1.91)$$

Where, H_t is transfer wave height behind the dike; H and L are height and wavelength of original wave in front of dike; B is dike top width.

The test results show that there is little difference with riprap embankment when workers block is used for surface protection.

The wave length behind the dike is determined by the shallow water wave length formula, usually assuming that the wave period is constant.

Chapter 3 The Interaction between Waves and Buildings

According to the standard of breakwater of Ministry of Transport, the curve in Fig. 3.1.20 is adopted to calculate the coefficient of transfer wave height K_t behind submerged embankment, and then the transfer wave height H_t is determined by pressing the formula below:

$$H_t = K_t H \qquad (3.1.92)$$

In Fig. 3.1.19, the horizontal coordinate is $(a - R_u)/H$. Where, a is the height of the top of the levee above the static tidal level, and the negative value is taken when the levee is submerged; R_u is the climbing height of the wave. When calculating R_u, it is considered as the case that the wave does not cross (assuming that a is large enough). This method was proposed by Davies et al.

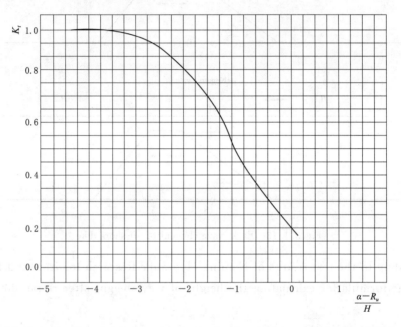

Fig. 3.1.19 Calculation diagram of wave height coefficient of submerged dike

2. Stability calculation of riprap submerged dike protection

Under the action of waves, riprap submerged dike protection block (or block) may lose stability, resulting in rolling, displacement and other damage. When the top of the submerged embankment is close to the static tidal level, the vulnerable parts are usually at the corner of the front slope shoulder and the top of the embankment. When the submerged depth of the embankment is large, the damage generally occurs at the top of the embankment and the front and rear slope shoulder. Another feature is that it is easier to reach the new stable state after certain damage occurs in submerged dike, and the damage degree is not easy to expand. Therefore, a slightly larger allowable failure rate is acceptable.

Nanjing Hydraulic Research Institute and Hohai University have obtained the follow-

3.1 The interaction between waves and sloping buildings

ing formula of stable mass of riprap (block) submerged levees:

$$W = \frac{K r_b H^3}{(S_r - 1)^3 m^b} K_\lambda K_a \quad (3.1.93)$$

$$b = \begin{cases} 1 & a/H \geqslant 1 \\ 0.24 \dfrac{a}{H} + 0.76 & a/H < 1 \end{cases} \quad (3.1.94)$$

$$K_a = \begin{cases} 1 & a/H \geqslant 1 \\ \left[\mathrm{ch}\left(1 - \dfrac{a}{H}\right)\right]^{-(0.8 + 0.05D)} & a/H < 1 \end{cases} \quad (3.1.95)$$

$$K_\lambda = \exp[4.5(0.05 - H/L)] \quad (3.1.96)$$

$$K = K_0 (1 + D)^{-0.33} \quad (3.1.97)$$

$$S_r = r_b / r$$

Where, a is the height of the top of the dike above the static tidal level; when emerging from the water, a takes a positive value; when submerging, a takes a negative value (Fig. 3.1.19); D is the allowable instability rate of the block; r_b and r are bulk density of stone and water, respectively; the coefficient b reflects the decreasing trend of the influence degree of m with the decreasing surface of the embankment top; K_λ reflects the influence of the wave steepness; K is the coefficient considering different blocks and different allowable instability rates; K_a is the stability parameter under the condition that the allowable instability rate $D = 0$; K_a reflects the influence of the water depth of the embankment top on the stability of the block; K_0 is listed in Table 3.1.14.

Table 3.1.14　　　　　　　　　　Stability parameter K_0

Block type	K_0	m
Block stone (two layers)	0.290	1.5 - 4.0
Block (two layers)	0.230	1.5 - 2.0
Quadrangular cone (two layers)	0.140	1.5 - 2.0
I-shaped block (two layers)	0.050	1.5
	0.057	2.0

According to the irregular wave test, Van der Meer et al. in the Netherlands believed that the stability of riprap under submerged embankment was mainly related to the relative embankment top height $(d + a)/d$ (d is the water depth) and damage level S. When $S = 2$, the damage begins, when $S = 5$, the damage is moderate, when $S = 8 - 12$, the damage is serious (exposed cushion and unacceptable). The relationship between the stability coefficient N_S and the relative embankment height S is as follows:

$$\frac{d + a}{d} = (2.1 + 0.1S) \exp(-0.14 N_S^*) \quad (3.1.98)$$

When $S = 2$:

Chapter 3 The Interaction between Waves and Buildings

$$N_S^* = 5.95 - \frac{1}{0.14} \ln \frac{d+a}{d} \tag{3.1.99}$$

According to the definition of Meer, the stable number N_S^* is:

$$N_S^* = \frac{H_S^{2/3} L_p^{1/3}}{\Delta D_{n50}} \tag{3.1.100}$$

$$\Delta = (\rho_a/\rho_w - 1)$$

$$D_{n50} = (W_{50}/\rho_a)^{1/3}$$

Where, H_S is the effective wave height; ρ_a, ρ_w are the density of the block and water; D_{n50} is the nominal diameter of the block or block; W_{50} is average quality.

Therefore, the formula of stable mass W of submerged embankment protection block or block can be obtained as follows:

$$W = \frac{r_b H_S^2 L_p}{\left(\frac{r_b}{r} - 1\right)^3 N_S^{*3}} \tag{3.1.101}$$

Where, L_p is the wavelength, m, corresponding to the spectral peak period T_p and it can be generally calculated as $T_{1/3} = 0.91 T_p$; N_S^* is determined according to Eq. (3.1.100) (corresponding initial damage, $S = 2$).

Meer's method is included in the *Code of Design for Breakwaters and Revetments* (JTS 154—2018) of Ministry of Communications.

3.2 The interaction of waves with a straight wall

3.2.1 Wave pattern in front of a straight wall

When a straight wall dike is built on the natural bank, the wave is reflected by the straight wall, and the wave in front of the dike is the interference wave after the incident wave and reflected wave superimpose. According to different wave elements, different water depth conditions and building scale factors (foundation bed conditions), wave forms in front of vertical walls can be generally divided into two categories: vertical wave and broken wave.

When the wave direction line is perpendicular to the straight wall, the scale of the straight wall along the crest line is greater than or equal to the wavelength, there is enough water depth in front of the wall, and the wall is smooth and impervious to water, complete reflection may be formed. The vertical wave is formed after the superposition of the reflected wave and the incident wave. Vertical waves are divided into deep water and shallow water according to the depth of water in front of the dike. If the wall is not smooth, or permeable holes and other reasons can not produce total reflection, the interference wave superimposed is incomplete vertical wave, or called partial vertical wave.

If the bottom water depth in front of the vertical wall is relatively large, but there is

3.2 The interaction of waves with a straight wall

a higher base bed under the wall, the waves transmitted to the base bed will be broken after sharp deformation due to the sharp reduction of the water depth. This kind of breaking will occur near the vertical wall, which is called near breaking wave.

If the sea water depth in front of the vertical wall is small, the base bed of the vertical wall is dark base bed or the base bed height is not large, then the amplitude of the interference wave is maximum at the abdominal point (distance from the embankment surface is $nL/2$), and the wave generally breaks here. For cases where wave breaking occurs at a distance greater than half a wavelength away from the wall, it is referred to as distant-breaking waves. Similarly, when there is no breakwater present, and the advancing waves break before reaching the location where a breakwater would be, this also falls under the category of distant-breaking waves.

According to the hydrological specifications for ports and waterways, the classification conditions of wave states in front of vertical walls are shown in Table 3.2.1.

Table 3.2.1   **Wave state division in front of vertical walls**

Base bed type	Conditions	Wave state
Dark and low base beds $\left(\dfrac{d_1}{d} > \dfrac{2}{3}\right)$	$\overline{T}\sqrt{g/d} < 8,\ d_1 \geqslant 2H$ $\overline{T}\sqrt{g/d} \geqslant 8,\ d_1 \geqslant 1.8H$	Vertical wave
	$\overline{T}\sqrt{g/d} < 8,\ d_1 < 2H,\ i \leqslant 1/10$ $\overline{T}\sqrt{g/d} \geqslant 8,\ d_1 < 1.8H,\ i \leqslant 1/10$	Far breaking wave
Medium base beds $\left(\dfrac{2}{3} \geqslant \dfrac{d_1}{d} > \dfrac{1}{3}\right)$	$d_1 \geqslant 1.8H$	Vertical wave
	$d_1 < 1.8H$	Near breaking wave
High bed beds $\left(\dfrac{d_1}{d} \leqslant \dfrac{1}{3}\right)$	$d_1 \geqslant 1.5H$	Vertical wave
	$d_1 < 1.5H$	Near breaking wave

In Table 3.2.1, H is the wave height where the building is located. According to the hydrographic regulations of the port channel, $H_{1\%}$ should be adopted, d is the water depth in front of the wall, d_1 is the water depth on the foundation bed, and i is the submarine slope ($i = \mathrm{th}\alpha$), as shown in Fig. 3.2.1. When there is a shoulder block on the plain base bed and the width is greater than 1.0 times the wave height, the water depth d_2 on the block is suitable to replace the water depth d_1 of the base bed to determine the wave state and wave force (Fig. 3.2.1).

When the steepness of the travelling wave is large $\left(\dfrac{H}{L} \geqslant \dfrac{1}{14}\right)$, the vertical wave formed in front of the dike may be the broken vertical wave with spray at the summit.

3.2.2 Vertical wave force

When the wave travels forward near the embankment, the water depth in front of the

Chapter 3   The Interaction between Waves and Buildings

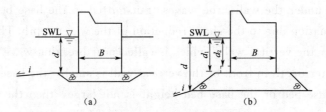

Fig. 3.2.1   The basic type of straight wall building

embankment is large enough, and the vertical wall is long enough, the embankment front will generate total reflection and form vertical wave. Vertical wave is a problem which has been studied earlier and further theoretically in terms of the interaction between waves and buildings. In 1928, French engineer Sainflou proposed the theoretical solution of vertical waves in shallow water, which is still used in engineering circles because of its simple calculation and sufficient accuracy in a certain range. Then Miche, Biesel, Rundgren and kyεHeoδ et al. proposed the second order approximate solution, and Tadjbakhsh and Keller, eagpagckaq proposed the third order approximate solution (the latter is actually the practical solution of Серекж-зеивкобии). Hong Guangwen (partial reflection of oblique wave) proposed a fourth order approximate solution.

1. Sainflou shallow water vertical wave

Using Lagrange method, Sainflou proposed the motion equation of a particle in shallow water vertical wave as:

$$\left. \begin{array}{l} x = x_0 + 2a \sin\sigma t \cos k x_0 \\ z = z_0 - 2b \sin\sigma t \sin k x_0 - 2kab \sin^2 \sigma t \end{array} \right\} \quad (3.2.1)$$

$$\left. \begin{array}{l} a = \dfrac{H}{2} \dfrac{\mathrm{ch} k(d-z_0)}{\mathrm{sh} kd} \\ b = \dfrac{H}{2} \dfrac{\mathrm{sh} k(d-z_0)}{\mathrm{sh} kd} \end{array} \right\} \quad (3.2.2)$$

Where, a and b are the semi-major and semi-minor axes of the water quality point motion ellipse corresponding to the original shallow water advancing wave; x_0 and z_0 are Lagrangian variables, x_0 and z_0 are taken as the position coordinates of the particle when it is at rest (rather than the position coordinates of the vibration center), the coordinate axes are taken in the static water surface, and the z axis is straight down.

Some of the following relationships for shallow water propulsion waves still apply:

$$\sigma^2 = kg \,\mathrm{th} kd$$
$$k = 2\pi/L$$
$$\sigma = 2\pi/T$$

That is, the vertical wave has the same wavelength and period as the original propulsive wave.

On the straight wall, the water particle can only have vertical displacement, so $x =$

3.2 The interaction of waves with a straight wall

x_0, $\cos kx_0 = 0$, $x_0 = \dfrac{L}{4}$, then the surface water quality point coordinates on the wall ($z_0 = 0$) is:

$$z = -2b\sin\sigma t - 2kab\sin^2\sigma t \qquad (3.2.3)$$

When $t = \dfrac{T}{4}$ and $t = \dfrac{3T}{4}$, $\sin\sigma t = \pm 1$, z takes the maximum and minimum values:

$$z_{\min}^{\max} = \mp 2b - 2kab \qquad (3.2.4)$$

It can be obtained that the maximum height of Sainflou's vertical wave is $|z_{\max} - z_{\min}| = 4b = 2H$, that is, the maximum height of the vertical wave is twice that of the original ongoing wave. The result is the same as that of micro wave vertical wave.

In the trajectory Eq. (3.2.1), the equation of wave surface curve at any depth in the case of vertical waves can be obtained by eliminating the parameter x_0:

$$z = z_0 - 2b\sin\sigma t \sin kx + 2kab\sin^2\sigma t \cos^2 kx \qquad (3.2.5)$$

According to Eq. (3.2.1), the height of the wave center line relative to the static tidal level is:

$$h_s = 2kab\sin^2\sigma t \qquad (3.2.6)$$

At water surface $b = b_0 = \dfrac{H}{2}$, and when $\sin\sigma t = 1$, the maximum elevation can be obtained:

$$h_s = 2ka_0 b_0 = \dfrac{\pi H^2}{L}\operatorname{cth} kd \qquad (3.2.7)$$

The height of the visible vertical wave is four times that of the original propulsive wave.

If t is eliminated from Eq. (3.2.1), the trace equation of water quality point of vertical wave in shallow water can be obtained:

$$z = z_0 - \dfrac{b}{a}(x - x_0)\tan kx_0 - \dfrac{kb(x - x_0)^2}{2a\cos^2 kx_0} \qquad (3.2.8)$$

It is known that the motion path of water quality point of vertical wave in shallow water is a parabola, and the principal axis of the parabola is the plumb line passing through the wave node ($x_0 = n\dfrac{L}{2}$, n is an integer).

The water quality point motion equation, trajectory and elevation of Sainflou wave are discussed above, and the vertical wave pressure calculation of engineering concern is discussed below.

In order to calculate the wave pressure of the vertical wave along the straight wall, the relevant partial derivative of Eq. (3.2.1) can be obtained and substituted into the Lagrange equation of motion. The second order terms $(ka)^2$ and $(kb)^2$ are omitted, and the third order terms (a/L) and (b/L) are obtained to obtain the pressure when any water quality point fluctuates.

Chapter 3 The Interaction between Waves and Buildings

$$\frac{p}{r} = z_0 + H \sin\sigma t \sin kx_0 \left[\frac{\mathrm{ch} k(d-z_0)}{\mathrm{ch} kd} - \frac{\mathrm{sh} k(d-z_0)}{\mathrm{sh} kd} \right] \quad (3.2.9)$$

On the straight wall, $x_0 = \frac{L}{4} + n\frac{L}{2}$ ($n = 0, 1, 2, \cdots$), $\sin kx_0 = \pm 1$, when there are peaks and troughs in front of the wall $t = \frac{T}{4}$ and $t = \frac{3T}{4}$, $\sin\sigma t = \pm 1$, then:

$$\frac{p}{r} = z_0 \pm H \left[\frac{\mathrm{ch} k(d-z_0)}{\mathrm{ch} kd} - \frac{\mathrm{sh} k(d-z_0)}{\mathrm{sh} kd} \right] \quad (3.2.10)$$

At the bottom of the wall, $z_0 = d$, the pressure is:

$$\frac{p}{r} = d \pm \frac{H}{\mathrm{ch} kd} = d \pm \frac{p_d}{r} \quad (3.2.11)$$

In the above equation, the former term is the hydrostatic pressure and the latter term $p_d/r = \pm H/(\mathrm{ch} kd)$ is the net wave pressure. It can be seen that the net wave pressure at the bottom of the water is twice the forward wave pressure.

Sainflou adopted Lagrange coordinates (x_0, z_0), and Eq. (3.2.10) and Eq. (3.2.11) represent the water quality point fluctuation pressure rather than the spatial point pressure. Therefore, when the total wave pressure of the straight wall is required, the wave pressure distribution of the straight wall should be drawn first. The calculation is rather complicated, and the simplified method of linear distribution is often adopted in practice. At the peak, the maximum net wave pressure is p; at the static tidal level, the underwater net wave pressure is p_d, and the peak pressure is zero. The pressure publication diagram with linear distribution is made according to these three points. During the trough, the maximum pressure fluctuation p'_s occurs at the bottom of the valley, the pressure at the static tidal level is zero, and the pressure at the bottom is p'_d. The pressure distribution diagram is also made according to these three points. The specific method is as follows:

(1) Peak time (Fig. 3.2.2). The fluctuating pressure p_s (kN/m²) at the static tidal level is:

$$p_s = (p_s + rd)\frac{H + h_s}{d + H + h_s} \quad (3.2.12)$$

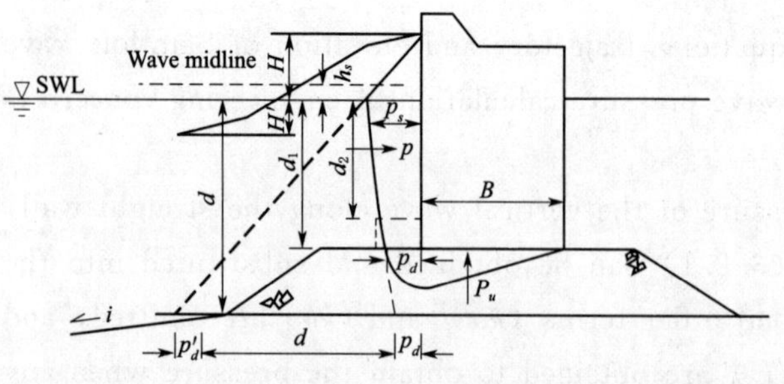

h_s is calculated according to Eq. (3.2.7).

The net wave pressure p_d (kN/m²) at the bottom is:

$$p_d = \frac{rH}{\mathrm{ch}\frac{2\pi d}{L}} \quad (3.2.13)$$

Fig. 3.2.2 Sainflou's vertical wave peak pressure diagram

Wave pressure at the bottom of the wall:

3.2 The interaction of waves with a straight wall

$$p_b = p_s - (p_s - p_d)\frac{d_1}{d} \quad (3.2.14)$$

The total wave force $p(\text{kN/m}^2)$ of the straight wall per unit length is:

$$p = \frac{1}{2}(H + h_s + d_1)(p_b + rd_1) - \frac{1}{2}rd_1^2 \quad (3.2.15)$$

The wave lifting force is equal to the lateral pressure at the front toe of the wall and zero at the rear heel. In the middle, the total wave lifting force $p_u(\text{kN/m}^2)$ acting on the bottom of the wall is:

$$p_u = \frac{1}{2}bp_b \quad (3.2.16)$$

(2) Trough time (Fig. 3.2.3). The total water pressure is less than the hydrostatic pressure at the trough, so the net fluctuation pressure is negative (in the opposite direction of the incoming wave).

The wave pressure at the bottom $p_d'(\text{kN/m}^2)$ is:

$$p_d' = \frac{rH}{\text{ch}\dfrac{2\pi d}{L}} \quad (3.2.17)$$

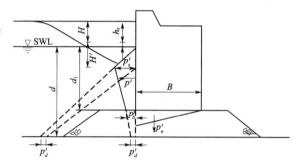

Fig. 3.2.3 Sainflou's vertical wave trough pressure diagram

The wave pressure of still water is zero, and the wave pressure strength $p_s'(\text{kN/m}^2)$ at the depth below still water $(H - h_s)$ is:

$$p_s' = r(H - h_s) \quad (3.2.18)$$

The wave pressure at the bottom of the wall $p_b'(\text{kN/m}^2)$ is:

$$p_b' = p_s' - (p_s' - p_d')\frac{d_1 + h_s - H}{d + h_s - H} \quad (3.2.19)$$

The total wave force $p'(\text{kN/m}^2)$ on the wall per unit length is:

$$p' = \frac{1}{2}rd_1^2 - \frac{1}{2}(d_1 + h_s - H)(rd_1 - p_b') \quad (3.2.20)$$

The wave lifting force on the bottom of the wall is negative (downward), and the strength of the front toe is equal to p_b'. Heel equals zero, total wave buoyancy $p_u'(\text{kN/m}^2)$.

$$p_u' = \frac{1}{2}p_b'b \quad (3.2.21)$$

The practice shows that the application range of Sainflou method is $d/L = 0.135 - 0.20$, wave steepness $H/L \geqslant 0.035$. When the relative water depth d/L is large, the result of this method is large, and when the relative water depth d/L is small, the result of this method is small. Therefore, according to the hydrological standard for port and channel, when wave steepness $H/L \geqslant 1/30$ and relative water depth $d/L = 0.1 - 0.2$, the

Chapter 3 The Interaction between Waves and Buildings

peak pressure of vertical wave shall be calculated according to Sainflou method. When the relative water depth is large, the wave crest is much larger by Sainflou method. When $d/L=0.5$, the calculated value is 1.3 – 3.0 times of the experimental value. Therefore, the standard stipulates that for $H/L \geqslant 1/30$, $0.5 > d/L > 0.2$, the vertical wave pressure at the peak should be calculated according to the first-order approximation of potential wave, while for the trough, the wave pressure should still be calculated according to Sainflou method.

For relative water depth $d/L \leqslant 0.12$ and $H/L = 1/20 - 1/30$, the peak pressure calculated by Sainflou method can be multiplied by the coefficient in Table 3.2.2, which is obtained by comparing the Miche-Bissel vertical wave method with Sainflou method.

Table 3.2.2 Correction factor of vertical wave pressure

d/L	$H/L=1/20$	$H/L=1/25$	$H/L=1/30$
0.08	1.33*	1.30	1.27
0.10	1.09	1.07	1.07
0.12	0.97	0.97*	0.97*

* Means the wave has broken in front of the wall.

For deep water ($d/L \geqslant 0.5$), $d = L/2$ can be used in calculation, and the wave compressive strength below $z = L/2$ can be set as zero.

For the calculation of wave pressure in the case of wave over the top, it is generally used to calculate the straight wall without going over the top first, and then deduct the wave force over the top in the wave pressure distribution diagram. This approach is biased towards safety.

The velocity of the front and bottom wall of Sainflou vertical wave can be determined as follows:

$$u_x = \frac{\partial x}{\partial t} - 2\sigma a \cos k x_0 \cos \sigma t \qquad (3.2.22)$$

For the water quality point at the wave belly at rest (n is an integer), $x_0 = \frac{L}{4} + n\frac{L}{2}$ (n is an integer), $\cos k x_0 = 0$, therefore $u_s = 0$; for the water quality point at the wave node at rest (n is an integer), $\cos k x_0 = \pm 1$, then:

$$u_x = 2\sigma a \cos \sigma t \qquad (3.2.23)$$

When the maximum crest and trough appear in front of the wall, then, all water quality points. When the wave surface passes the static tidal level, $\cos \sigma t = \pm 1$, then:

$$u_x = 2\sigma a \cos k x_0 \qquad (3.2.24)$$

When the wave surface passes through the static tidal level, the maximum value of the horizontal sub-velocity is taken:

3.2 The interaction of waves with a straight wall

$$u_{x,\max} = \pm 2\sigma a \qquad (3.2.25)$$

Compared with the forward wave, the maximum horizontal sub-velocity of the vertical wave is twice that of the forward wave.

In the water bottom $z_0 = d$, the horizontal bottom velocity of the water quality point is:

$$u_d = \frac{2\pi H}{\sqrt{\dfrac{\pi L}{g} \mathrm{sh} 2kd}} \cos kx_0 \cos \sigma t \qquad (3.2.26)$$

The maximum bottom velocity occurs when the wave surface passes through the static tidal level and at the nodal position:

$$u_d = \frac{2\pi H}{\sqrt{\dfrac{\pi L}{g} \mathrm{sh} 2kd}} \qquad (3.2.27)$$

Sainflou method for calculating vertical wave pressure has some shortcomings. Except as previously mentioned, this method tends to overestimate when the relative water depth d/L is large, and underestimate when d/L is small. the test also confirms that:

1) The process line of Sainflou wave pressure and the process line of the wall front change in the same law, that is, the law of maximum pressure at the peak and minimum pressure at the trough is not completely consistent with the reality. When the wave steepness is large, the wave pressure process line appears saddle shape, and the maximum wave pressure does not appear at the peak moment. At the same time, when the front wave surface of the wall passes through the static tidal level, the net wave pressure of each point on the wall is not equal to zero.

2) In the case of deep water, due to saddle shape of steep wave pressure process line, negative wave pressure may appear in the lower part of the wave pressure distribution diagram, while Sainflou vertical wave pressure is always positive or tends to zero below deep water, and no negative pressure will appear.

In addition to theoretical discussion by many scholars, a large number of laboratory experiments and field observations have also been conducted to study vertical wave pressure, and some empirical methods have been proposed, such as Nagai Shoshichiro, Goda Yoshimi of Japan.

Goda Yoshimi proposed a new formula for calculating both vertical and broken waves, taking into account irregular and oblique wave actions. The method has been included in the technical standard of Japanese port facilities.

$H_{\max}$, the maximum wave height of the design wave, is taken as $H_{\max} = 1.8 H_{1/3}$ outside the breaking zone. In the breaking zone, $H_{\max}$ is taken as the distance from $5H_{1/3}$ in the offshore direction in front of the breakwater, and the latter is calculated according to the deformation in the breaking zone. At this time, the water depth of H_S is determined by the water depth at the breakwater.

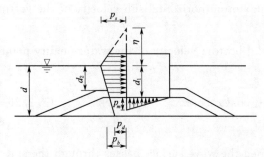

Fig. 3.2.4 The Aida wave pressure diagram

Maximum wave period $T_{max}=T_{1/3}$.

The wave pressure acting height η (Fig. 3.2.4) is:

$$\eta=0.75(1+\cos\beta)H_{max} \qquad (3.2.28)$$

Where, β is the included angle between the vertical line of the breakwater wall and the main wave direction. When designing, the instability of wave direction should be considered, with the main wave direction swinging 15° towards the hazardous side. The intensity of wave pressure against a vertical wall surface:

$$p_1=\frac{1}{2}(1+\cos\beta)(a_1+a_2\cos^2\beta)rH_{max} \qquad (3.2.29)$$

$$p_2=\frac{p_1}{\text{ch}\dfrac{2\pi h}{L}} \qquad (3.2.30)$$

$$p_s=a_3 p_1 \qquad (3.2.31)$$

Where

$$a_1=0.6+\frac{1}{2}\left[\frac{4\pi d/L}{\text{sh}(4\pi d/L)}\right]^2 \qquad (3.2.32)$$

$$a_2=\min\left\{\frac{d_3-d_2}{3d_3}\frac{H_{max}}{d_2},\frac{2d_2}{H_{max}}\right\} \qquad (3.2.33)$$

$$a_3=1-\frac{d_1}{d}\left[-\frac{1}{\text{ch}(2\pi d/L)}\right] \qquad (3.2.34)$$

Where, $\min\{a,b\}$ is the minimum of a and b; d_3 is the water depth at $5H_{1/3}$ on the seawall side from the breakwater wall.

When the top of the breakwater is low and waves are generated, the above wave pressure distribution diagram remains unchanged, and the waves are deducted when calculating the total pressure.

The wave lifting force is assumed to be triangular in distribution, with the strength of the front toe p_u and the heel zero:

$$p_u=\frac{1}{2}(1+\cos\beta)a_1 a_3 rH_{max} \qquad (3.2.35)$$

The total wave force can be calculated according to the wave pressure distribution diagram.

The Aida method is only used in the case of crest action. Because the formula includes the calculation of the wave pressure of the vertical wave and the broking wave, the discontinuity phenomenon during the transition from the vertical wave to the breaking wave is avoided in the previous calculation of the wave pressure.

3.2.3 Far breaking wave force

1. *Code of Hydrology for Harbour and Waterway* (JTS 145—2015)

The study shows that the pressure of far breaking wave is not only related to wave elements (e. g. wave height, wave steepness etc.), but also closely related to the bottom slope. The steeper the bottom slope, the higher the pressure increases. The test result of Dalian University of Technology is adopted in the *Code of Hydrology for Harbour and Waterway* (JTS 145—2015).

(1) Peak time (Fig. 3.2.5).

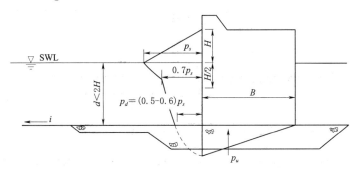

Fig. 3.2.5 Wave pressure diagram of far breaking wave

The wave pressure intensity at H above the still water surface is zero.

The wave pressure strength at static water surface p_s (kN/m²) is:

$$p_s = r K_1 K_2 H \tag{3.2.36}$$

Where, K_1 is related to the bottom slope, according to Table 3.2.3; K_2 is related to the wave steepness, according to Table 3.2.4.

Table 3.2.3　　　　　　　　　　　　Coefficient K_1

Bottom slope i	1/10	1/25	1/40	1/50	1/60	1/80	1/100
K_1	1.89	1.54	1.40	1.37	1.33	1.29	1.25

Note: The bottom slope i is the mean value within a certain distance in front of the building.

Table 3.2.4　　　　　　　　　　　　Coefficient K_2

Wave steepness L/H	14	15	16	17	18	19	20	21	22
K_2	1.01	1.06	1.12	1.17	1.21	1.26	1.30	1.34	1.37
Wave steepness L/H	23	24	25	26	27	28	29	30	
K_2	1.41	1.44	1.46	1.49	1.50	1.52	1.54	1.55	

The wave pressure strength above the static water surface changes in a straight line. The wave pressure $p_z = 0.7 p_s$ at the depth $z = H/2$ below the static water surface, and the wave pressure p_d at the bottom. When d/H is 1.7, $p_d = 0.6 p_s$; When d/H is

1.7, $p_d = 0.5 p_s$.

The total buoyancy force p_u on the bottom of the wall is:

$$p_u = \mu \frac{b p_d}{2} \tag{3.2.37}$$

Where, μ is the reduction coefficient of floating support force (also known as saturation coefficient).

(2) Wave trough (Fig. 3.2.6).

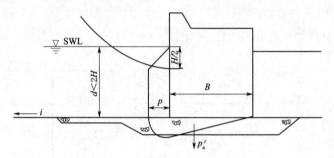

Fig. 3.2.6 Wave trough pressure diagram of far breaking wave

The far breaking wave pressure at the trough can be calculated as follows. The wave pressure strength is zero at still water. The wave pressure intensity p is constant from the depth $z = H/2$ below the still water surface to the bottom.

$$p = 0.5 \gamma H \tag{3.2.38}$$

When the bottom slope is relatively flat ($i = 1/50$), the limit wave height of traveling wave (generally $H = 0.78 d$) can be used to calculate the far-breaking wave force according to the above method, where the design wave has broken at or beyond the breakwater before it is built. For steeper bottom slope ($i > 1/50$), larger wave force may be generated. In this case, the limit wave height can be preliminarily estimated according to the above far breaking wave pressure, and then verified by model test.

For a vertical wall with a low base bed, the pressure distribution of the far breaking wave can be plotted according to the water depth d in front of the building, and then the wave force of the base bed part can be subtracted.

2. The former Soviet Code СНиП I

The former Soviet Code СНиП I proposes a set of calculation methods for determining the wave pressure of straight wall buildings located in the water boundary zone. It is calculated in the following three cases.

(1) When the building is arranged at the last breaking line of the shore wave [Fig. 3.2.7 (a)]:

$$p_1 = \gamma H_b \left(0.033 \frac{L}{d} + 0.75 \right) \tag{3.2.39}$$

$$\eta_1 = \frac{p_1}{\gamma} \tag{3.2.40}$$

(2) When the building is arranged within the water edge [Fig. 3.2.7 (b)]:

$$p_2 = \left(1 - 0.3\frac{l_2}{l_1}\right) p_1 \quad (3.2.41)$$

$$\eta_2 = \frac{p_2}{\gamma} \quad (3.2.42)$$

(3) When the building is arranged in the wave climbing range on the bank slope behind the water edge [Fig. 3.2.7 (c)]:

$$p_3 = 0.7\left(1 - \frac{l_3}{l_4}\right) p_1 \quad (3.2.43)$$

$$\eta_3 = \frac{p_3}{\gamma} \quad (3.2.44)$$

Where, η_1, η_2, η_3 are the crest height above the tidal level calculated before the straight wall; H_b is wave breaking height; l_1 is the distance from the final breaking line of the wave to the water edge line; l_2 is the distance from the last breaking line of the wave to the building; l_3 is the distance from water edge to land side buildings; l_4 is the distance from the water edge to the upper limit of the wave's climb on the bank slope (without buildings).

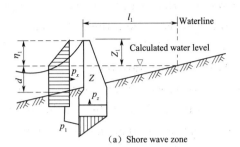
(a) Shore wave zone

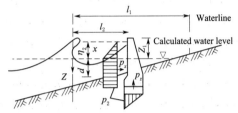
(b) Water boundary area

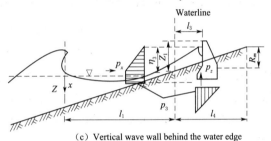
(c) Vertical wave wall behind the water edge

Fig. 3.2.7 Wave trough pressure diagram of the building

The above method is suitable for the calculation of wave pressure in the water boundary region (the last wave breaking and climbing region). It is applicable to the condition that the height of the wall top is not lower than $0.3H_b$ above the static tidal level.

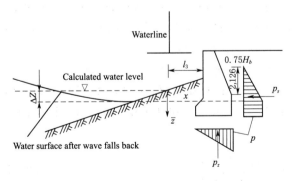

Fig. 3.2.8 Wave pressure distribution on vertical protective wall when wave falls back

For a wave wall with backfill on the land side (Fig. 3.2.8), the pressure at the trough (when the horizontal force is seaway) should be calculated as follows. The pressure p from ΔZ below the static tidal level to the bottom of the wall is:

$$p = \gamma(0.75H_b + \Delta Z) \quad (3.2.45)$$

In the equation, ΔZ is the decrease

value of water surface when the wave falls back in front of the straight wall, which is related to the distance l_3 from the water edge to the building. When $l_3 \geqslant 3H$, $\Delta Z = 0$, when $l_3 < 3H_b$, $\Delta Z = 0.25H_b$.

3.2.4 Near breaking wave force

Near breaking wave refers to the situation that the wave is broken directly in front of the wall or on the wall, and the near broken wave generates huge wave force, which generally occurs in the case that the base bed of the mixed dike is high. In this case, the water depth on the beach in front of the dike is large, and the wave is not broken, because when the wave moves to the base bed, the water depth suddenly decreases and the wave breaks, forming a huge impact on the wall. The main factors affecting the near breaking wave force are wave element, water depth in front of the wall and embankment, scale factor of the foundation bed (width of foundation shoulder, slope degree in front of foundation bed and roughness characteristics, etc.). Choosing the main factor, the total wave force (horizontal force) acting on the straight wall can be written as the following dimensionless relation:

$$\frac{p}{\gamma H^2} = f\left(\frac{d_1}{H}, \frac{L}{H}, \frac{b}{H}, \frac{d}{H}, m\right) \tag{3.2.46}$$

Where, p is the total horizontal force, kN; H is the calculated wave height, m; d_1 is the water depth in front of the wall, m; d is the water depth in front of the dike, m; b is the width of the foundation shoulder, m; m is the slope ratio in front of the foundation bed.

The existing methods for calculating breaking wave pressure include Guangjing Yong, Minikin, Pulakida, Dalian University of Technology, etc. According to Hohai University, the wave pressure calculation method of the top wall of slope dike also includes the case of near breaking wave when the water depth in front of the wall is large.

The study on the wave pressure of the straight wall with high foundation bed shows that the water depth in front of the wall has the most important influence on the breaking wave pressure. Generally, the maximum pressure occurs within the range of $d_1/H = 0.5 - 0.9$. For different wave steepness H/L and relative shoulder width b/H, the water depth in front of the wall with the maximum pressure d_1/H and the value of the maximum pressure are different, presenting a very complex staggered state.

1. *Code of Hydrology for Harbour and Waterway* (JTS 145—2015)

The *Code of Hydrology for Harbour and Waterway* (JTS 145—2015) adopts the method based on experiment obtained by Dalian University of Technology. The dimensionless total wave force is written as:

$$\frac{p}{\gamma H d_1} = f\left(\frac{d_1}{d}, \frac{H}{d_1}, \frac{H}{L}\right) \tag{3.2.47}$$

According to the test results, it is concluded that the wave force of near breaking

3.2 The interaction of waves with a straight wall

wave has little relationship with wave steepness H/L. When the shoulder width is relatively narrow, the wave deformation is not obvious, and the steep wave is easy to break and form a large impact force; when the shoulder width is relatively large, the wave deformation is sufficient, and the steep slope breaks earlier, and the wave force is relatively large due to the large energy of the tempera wave. Therefore, the steepness factor of the wave is omitted, and only the influence of d_1/d and H/d_1 is considered. This method is only applicable to the condition of $d_1 \geqslant 0.6H$.

At the wave crest, the wave intensity at the height $Z(m)$ above the still water surface is zero, and Z is calculated according to the formula below:

$$Z = \left(0.27 + 0.53 \frac{d_1}{H}\right) H \tag{3.2.48}$$

Static water wave pressure strength $p (kN/m^2)$, divided into the following two cases:

(1) $\frac{1}{3} < \frac{d_1}{d} \leqslant \frac{2}{3}$ (medium base bed):

$$p_z = 1.25 \gamma H \left(1.8 \frac{H}{d_1} - 0.16\right)\left(1 - 0.13 \frac{H}{d_1}\right) \tag{3.2.49}$$

(2) $\frac{1}{4} \leqslant \frac{d_1}{d} \leqslant \frac{1}{3}$ (high base bed):

$$p_z = 1.25 \gamma H \left[\left(13.9 - 36.4 \frac{d_1}{d}\right)\left(\frac{H}{d_1} - 0.67\right) + 1.03\right]\left(1 - 0.13 \frac{H}{d_1}\right) \tag{3.2.50}$$

The strength of the wave pressure at the bottom of the wall:

$$p_b = 0.6 p_s \tag{3.2.51}$$

The total wave force on the wall per unit length can be calculated under the following two conditions:

(1) $\frac{1}{3} < \frac{d_1}{d} \leqslant \frac{2}{3}$ (medium base bed):

$$p = 1.25 \gamma H \left(1.9 \frac{H}{d_1} - 0.17\right) \tag{3.2.52}$$

(2) $\frac{1}{4} \leqslant \frac{d_1}{d} \leqslant \frac{1}{3}$ (high base bed):

$$p = 1.25 \gamma H d_1 \left[\left(14.8 - 38.8 \frac{d_1}{d}\right)\left(\frac{H}{d_1} - 0.67\right) + 1.1\right] \tag{3.2.53}$$

Wave lifting force on the underside of a wall:

$$p_u = \mu \frac{b p_b}{2} \tag{3.2.54}$$

Where, μ is the reduction coefficient of wave lifting force distribution diagram which can be 0.7.

See Fig. 3.2.9 for the wave pressure diagram.

Chapter 3 The Interaction between Waves and Buildings

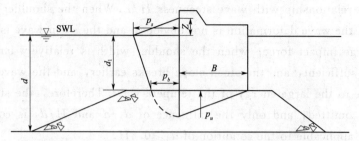

Fig. 3.2.9 Wave pressure diagram of near breaking wave

2. Minikin method

The calculation method of breaking wave pressure proposed by Minikin in 1950 is still used in the west (such as SPM in 1984). According to this method, the pressure is composed of two parts: the dynamic water pressure acting on the area near the static tidal level and the hydrostatic pressure acting on the whole embankment surface (Fig. 3.2.10).

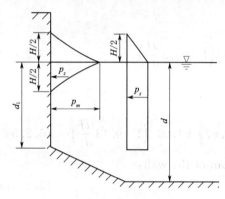

Fig. 3.2.10 Minikin breaking wave pressure

(1) Dynamic water pressure: acting in the range of $H/2$ above and below the static tidal level, it is a parabolic distribution. At the static water surface, it is:

$$p_m = 101\gamma d_1 \left(1 + \frac{d_1}{d}\right)\frac{H}{L} \qquad (3.2.55)$$

The dynamic pressure decreases from the static tidal level up and down and reaches zero at $H/2$, z is from the static tidal level:

$$p_z = p_m \left(1 - \frac{2|z|}{H}\right)^2 \qquad (3.2.56)$$

(2) Additional hydrostatic pressure section. Above static tidal level (z from static water level):

$$p_s = \gamma \left(\frac{H}{2} - z\right) \qquad (3.2.57)$$

Below static tidal level:

$$p_s = \frac{1}{2}\gamma H \qquad (3.2.58)$$

The calculation methods of wave breaking pressure by *Code of Hydrology for Harbour and Waterway* (JTS 145—2015) and Minikin are introduced above. The *Code of Hydrology for Harbour and Waterway* (JTS 145—2015) does not consider the effect of wave steepness, which is a problem worthy of further study and discussion. Minikin's formula, though including the effect of wavelength, has been simply considered to be inversely proportional to wavelength. This result has also been controversial. It is noteworthy that Aida's method reflects the existence of periodic effects and this effect is evident

3.2 The interaction of waves with a straight wall

from the results of wave pressure tests and breakwater model sliding tests. According to the experiments of Hohai University in recent years, the influence of wave steepness obviously exists, but the variation law is complicated.

3.2.5 Oblique wave action

The above calculation methods of wave pressure for wave interaction with straight wall refer to the positive wave action, but in fact the wave action is often oblique. At present, in the case of unbroken wave, there are some theoretical answers (e.g. Hong, 1974) in the study of oblique wave force, and other cases are mostly dealt with some empirical methods.

The reflection test of the vertical dike find that the following three situations occurred with the difference of the incident angle θ_i between the incoming wave line and the vertical dike wall normal direction.

(1) When the incident angle $\theta_i \leqslant 45°$, it is normal reflection, and the incident angle equals the reflection angle, that is, the reflected wave completely leaves the dike [Fig. 3.2.11 (a)].

(2) When $45° < \theta_i < 70°$, a Mach reflection occurs like in acoustics. The wave crest line at the dike body becomes perpendicular to the dike body [Fig. 3.2.11 (b)], forming a stem that advances and grows along the dike. The wave crest outside the stem advances along the reflected wave line, so the wave crest line becomes an outward-convex broken line along the wave direction, and the reflection angle θ_r is less than incidence angle θ_i.

(3) When $\theta_i \geqslant 70°$, no reflection occurs. The crest line of the incident wave turns in front of the dike and still forms a stem near the wall. The crest line forms a concave fold line [Fig. 3.2.11 (c)].

According to the experiment and theoretical deduction, the relationship between stem angle ρ and incidence angle θ_r, θ_i is as follows:

$$\rho = \theta_i - \theta_r \qquad (3.2.59)$$

Where, ρ obtained from the test is roughly as follows: when $\theta_i = 45°$, $\rho = 1°$; when $\theta_i = 80°$, $\rho = 10°$.

The force of oblique wave depends on the size of the interference wave in front of the dike. Since the interference wave surface changes along the wall, the total wave force along the single wide wall of the dike line also changes. It is observed that the interference wave height is higher than 2 times incident wave height in some places, and less than 2 times incident wave height in other places.

On the basis of the test results, the 24th International Shipping Conference proposed the following equation to calcu-

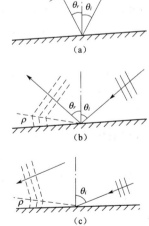

Fig. 3.2.11 Oblique wave reflection

late the interference wave high H_f:

$$H_f = H_i(1+\cos^{1/2}\theta_i) \quad (3.2.60)$$

Where, H_f is incident wave height; θ_i is the angle between the incident wave direction and the normal direction of the wall, which is the forward effect when $\theta_i=0°$. The above formula applies to: when $d/L \leqslant 0.1$, $0° \leqslant \theta_i \leqslant 90°$; when $d/L \geqslant 0.1$, $\theta > 60°$.

At that time, it is the case of complete reflection under forward action, and $H_f = 2H_i$, so the interference wave height of oblique wave is less than that of total reflection wave under forward action. According to Eq. (3.2.60), in the case of oblique wave, relative to the height of the completely reflected vertical wave, the reduction coefficient of the interference wave height is:

$$K_H = \frac{H_f}{2H_i} = 1/2(1+\cos^{1/2}\theta_i) \quad (3.2.61)$$

In the range of the first order wave, the wave pressure strength is linearly related to the wave height, so the total wave force reduction coefficient of the oblique wave is K.

According to Hong's theoretical solution, the total wave force reduction coefficient of the oblique wave is related not only to the incident angle θ_i, but also to the relative water depth d/L and the relative length of the straight wall section l/L. According to the analysis, when $d/L = 0.15 - 0.2$ and $l/L < 0.3$, the calculation result is similar to Eq. (3.2.61).

The above discussion refers to the case where the waves do not break. Some foreign standards also stipulate the total wave force reduction coefficient of the breaking wave and the broken wave, such as SPM of the United States and CH 92-60 of the former Soviet Union, etc. $K_p = \cos^2\theta_i$ (here θ_i is the angle between the incident wave and the wall normal).

In terms of the combined pressure on the straight wall section under the action of the inclined wave, the experimental and theoretical results show that the action of the inclined wave may be larger than that under the forward wave in terms of the maximum point pressure.

At the same time, it must be pointed out that the erosion caused by the velocity of the front bottom of the wall is paid more and more attention by researchers and engineers.

Exercise

Q1 Vocabulary explanation: Vertical wave, near-breaking wave, far-breaking wave.

Q2 Fill in the blanks: The types of foundation beds of straight wall buildings include ( ), ( ), ( ), and the criterion is ( ).

Q3 Multiple choices: The methods that can be used to calculate the vertical wave force are ( ).

Exercise

A. Shallow water vertical wave method based on elliptic cosine wave

B. Sempro reduction method

C. Yoshimi Goda method

D. Minikin method

Q4 Fill in the blanks: When calculating the vertical wave force, the formula of (   ) can only calculate the peak force, while the formula of (   ) is generally used to calculate the trough force. In the layout of the breakwater and the development of the structural form, should try to avoid the (   ) wave at the (   ) tidal level.

Q5 Short answer: What are the different effects of waves on a building with a straight wall when they arrive at an oblique incident angle of 30° and 10° respectively?

Q6 Fill in the blanks: The advantages and disadvantages of the angular vertical dike are (   ) and (   ).

Q7 Short answer: What are the morphological characteristics of sand scouring in front of vertical dike under vertical wave and what are the related factors?

Q8 Fill in the blanks: The influence range of wave on the vertical embankment is mainly (   ), and the maximum scouring depth increases with the length of protective layer (   ).

Q9 Vocabulary explanation: Climb high, fall deep.

Q10 Fill in the blanks: The main influencing factors of climbing height include (   ), (   ), (   ), (   ), wave scale, etc.

Q11 Vocabulary explanation: Average wave overtopping, single wave overtopping.

Q12 Fill in the blanks: The main influencing factors of wave overtopping are (   ), (   ), (   ), (   ), slope of building bottom, etc.

Q13 Fill in the blanks: The main contents of slope protection design include: (   ), (   ), (   ), (   ).

Q14 Short answer: What are the instability modes of the armour block?

Q15 Fill in the blanks: Hudson formula is proposed based on the instability pattern of (   ), Ilibarren formula is proposed based on the instability pattern of (   ), and Chaitsev formula is proposed based on the instability pattern of (   ).

Q16 Multiple choices: What factors is the stability factor mainly related to? (   )

A. Block type   B. Throwing mode

C. Excess wave quantity   D. Allowable instability rate

Q17 Vocabulary explanation: Allowable instability rate.

Q18 Choices: What are the blocks with an allowable instability rate of 0? (   )

A. Throw and fill 2-layer block stone   B. Twist Wang block

C. Twist I - shaped block   D. Square

Q19 Short answer: What are the similarities and shortcomings of Ilibarren formula, Hudson formula and Chaitsev formula?

Chapter 3 The Interaction between Waves and Buildings

Q20 Multiple choices: Which of the following is true? (   )

A. When the top elevation of the underwater stone throwing edge is 1.0 times the design wave height below the design low tidal level, the weight of the stone throwing edge is smaller than the stable weight of the slope protection block.

B. The weight of the outer slope protection cushion block is less than the weight of the protection block.

C. The weight of the dike top block is always the same as that of the outer slope block at any time.

D. The weight of the embankment head protective block should be less than that of the embankment body protective block.

Q21 Choices: What is incorrect statement about the stable weight of the inner slope protection block? (   )

A. When the top elevation of the dike is not determined according to the wave overtopping, the inner slope protection block should be made of block stone with the same weight as the outer slope protection cushion.

B. When the top elevation of the dike is determined according to the wave overtopping, the weight of the inner slope protection block between the top of the dike and the design low tidal level below 0.5 - 1.0 times the design wave height shall be the same as that of the outer slope protection block.

C. When the top elevation of the dike is determined according to the wave overtopping, the inner slope protection block under the design low tidal level is 0.5 - 1.0 times the design wave height, and the same weight as the outer slope protection cushion should be used, but should not be less than 150 kg, and should be checked according to the wave inside the dike.

D. When the dike top elevation is determined according to no wave overtopping, the weight of the inner slope protection block should not be less than the weight of the outer slope protection cushion block stone.

Q22 Fill in the blanks: Concrete slope protection design includes (   ) and (   ).

Q23 Multiple choices: The influencing factors of wave lifting force are (   ).

A. Front wave elements B. Wave climb
C. Top elevation D. Tidal level changes inside and outside the dike

Q24 Multiple choices: The main factors influencing the wave action on the parapet at the top of slope are (   ).

A. Stability coefficient of the protective block

B. Slope grade

C. Water depth in front of the parapet

D. Rough permeability characteristics of slope surface

Q25 Multiple choices: The main influencing factors of the far-breaking wave force are (   ).

    A. Wave element   B. Slope of base bed

    C. Roughness of base bed   D. The relative depth of water on the base bed

Q26 Multiple choices: The main influencing factors of the near breaking wave force are (   ).

    A. Relative width of base shoulder   B. Slope of base bed

    C. Wave elements   D. The relative depth of water on the base bed

Chapter 4 Seadike Engineering

A seadike is a kind of dike built along the coast to mitigate tides and waves and prevent inundation, and is the main facility of coastal protection engineering. According to the *Hydraulic Dictionary*, the coastal protection engineering is defined as "a shore protection project that protects coastal areas against erosion and scouring by sea-swells, alongshore currents and tidal currents, as well as flooding and inundation induced by storm surges. Mainly includes seadike, shore protection and beach preservation". Therefore, the construction of a seadike is a type of coastal protection project, and can also be a part of the artificial island project or reclamation engineering. Seadike projects are characterized by the requirements for the mitigation of waves and tides, soft foundations and the need to be conducted mostly underwater, comparing to common channel structures. In some coastal areas in China, the seadike is also known as a sea pond.

4.1 Management and layout of the dike line

The following principles should be followed in the management and layout of the seadike.

(1) The planning of embankment lines should comply with the unified planning and comprehensive development and utilization needs of the coastal zone, taking into account reclamation, aquaculture, flood control and drainage, shipping and transportation, and requirements for ship shelter. In estuarine areas, it should also be integrated with local estuary management and planning measures.

(2) Environmental and ecological protection should be emphasized. An appropriate environmental assessment is needed. Negative impacts in neighboring areas and industry departments should be avoided. Possible environmental problems and their responses should be presented.

(3) The dike line layout should generally be compared with multiple options. According to the actual local conditions, the morphology, geology, hydrology, building materials and construction conditions should be considered. The option with less investment, more convenient construction and better overall efficiency should be selected.

(4) The layout of the dike line shall be straight to avoid excessive curvatures, con-

cave and convexity which can lead to localized wave energy concentration. Meanwhile, it is important to choose an alignment that is favorable to wave protection and to avoid dike lines that are orthogonal to the direction of large waves. However, straightening the dike line and increasing the amount of work are not necessary.

(5) The flat elevation at the dike line should be determined according to the purpose of the dike construction, development approach, dike safety, construction conditions, etc. For dikes as agricultural polder, it is inappropriate to simply pursue the large area enclosed by unit dike line, which would lead to low flat elevations, poor foundations and increased difficulties in constructions. The flat elevation on mudflats suitable for foothill farming should be considered to facilitate the conservation and development of farming resources.

4.2 Types of cross-shore section of seadike

Types of the cross-shore section of seadikes are distinguished by their waterfront shape characteristics. There are three types of seadikes: sloping seadike, vertical (including upright) seadike and composite seadike.

4.2.1 Sloping seadike

The waterfront slope of the seadike is usually set as $m = \cot\alpha > 1$. The body of the seadike is mainly filled withsoil material, with an armour layer equipped on the waterfront slope of the seadike. The types of armour layer include dry blocks or slates, slurry blocks, ripraps, precast concrete panels, cast in-situ integral concrete, asphalt concrete, artificial blocks, cemented soil and turf cover etc. The waterfront slope of the sloping seadike is mild, implying a good stability and weak wave reflection. The wide base of the seadike allows for a more even distribution of foundation stress, making it more advantageous to build on the mudflats. Sloping seadikes are easily constructed, locally sourced, adaptable to wind and wave induced deformation and local damages, and easy to repair. However, the main disadvantages of the sloping seadike include its large cross-shore section, which requires more works and covers a larger area. Larger wave run-up could occur on the waterfront slope within a certain slope range. Since seadike construction often requires mounding first and armouring later, constructed earthworks are likely to be washed away at low tidal flat. Therefore, sloping seadikes are generally used for projects in high-level reclamation above neap high tidal levels.

The typical sloping seadike is depicted in Fig. 4.2.1.

The waterfront slope of the sloping seadike can also be classified into three types: single sloping type, folded sloping type and composite sloping type. The single sloping type means that there is only one slope from top to bottom, as can be seen in Fig. 4.2.1. The folded sloping type is a slope with an inflection point and two different slopes above and below the inflection point, as shown in Fig. 4.2.2.

Chapter 4　Seadike Engineering

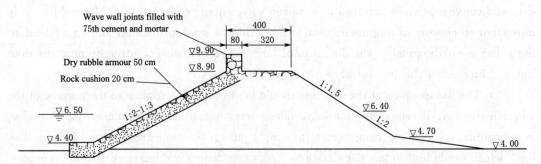

Fig. 4.2.1　Sloping seadike in front of Xiangshanmen (unit in length: m, unit in elevation: m)

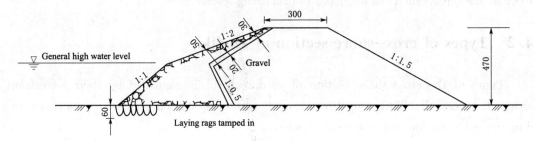

Fig. 4.2.2　Sloping seadike with two different slopes (unit: cm)

The composite sloping seadike is formed with a platform at a certain elevation on the slope, as can be seen in Fig. 4.2.3.

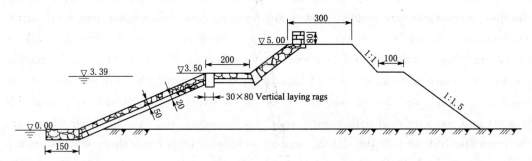

Fig. 4.2.3　Composite sloping seadike at Nanbei Yang in Putian
(unit in elevation: m, scale: cm)

The elevation of the platform should be set at the high tide. The platform not only contributes to the stability of the seadike, but also has a significant effect on reducing wave run-up on the slope. The concentration of wave energy at the corners near the platform needs to be strengthened with appropriate protective actions.

In order to reduce the volume of earth and stone required for the cross-section of the embankment without compromising the standard for resisting wave overtopping, a wave barrier parapet approximately 1.0 m high is often installed on the crest of the embankment. See Fig. 4.2.1. To prevent the embankment toe from being eroded by tides and waves, which could compromise its stability, it is common to install a rubble mound revetment at the toe of the embankment, or construct a large apron or rubble masonry. Other methods include placing single or double piles with stone inlays at the front slope. If the foundation is soft and the load on the seadike body is large, to maintain the stability of the seadike foundation, it is necessary to make a ballast layer at the bottom of both sides of the seadike or lay a drainage sand bedding layer on the foundation, etc., as can be seen in Fig. 4.2.4.

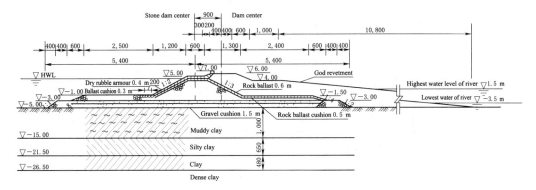

Fig. 4.2.4 Section of the vertical seadike at the shallow part of Hucheng Port
(unit in scale: cm, elevation: m)

4.2.2 Vertical seadikes

The vertical seadikes arethe most common around Zhoushan, Zhejiang Province. The front slope of the vertical seadike is built with blocks or slates which form a steep wall of $m<1$. A backfilter layer or riprap slag is set up behind the wall, and then the backfilling is conducted. As can be seen in Fig. 4.2.5, the width of the crest of the stone wall is about 0.5 – 1.5 m, with every stone weighting about 50 – 100 kg at the outer layer of the waterfront slope. The vertical seadike has a smaller cross-shore section, which can reduce the amount of filling works. It is recommended that the stone works should be conducted first to cover the filling materials with stones, which can reduce the loss of the filling material during the construction. Therefore, it can be used for the dike projects at neap low tides as well as the low tidal flats. Compared with the sloping seadike, the foundation stress is more concentrated under the vertical seadike, promoting a larger and more concentrated sink of the dike body. Hence, the requirements for the foundation of the vertical seadike should be higher. In some areas, stone walls are often built on a foundation bed because of the low loading capacity of the muddy foundation or large water depths. From the perspective of wave actions, the wave reflection in front of a steep seadike is large and the wave types are standing waves or incompletely standing waves. Therefore, the wave runup on the vertical seadike is generally smaller than that on a sloping seadike. However, it is important to note that the increased bottom velocity caused by standing waves can lead to scouring at the toe of the seadike. In addition, wave actions are intense and beating the seadike, with water splashing and causing more damage to the seadike.

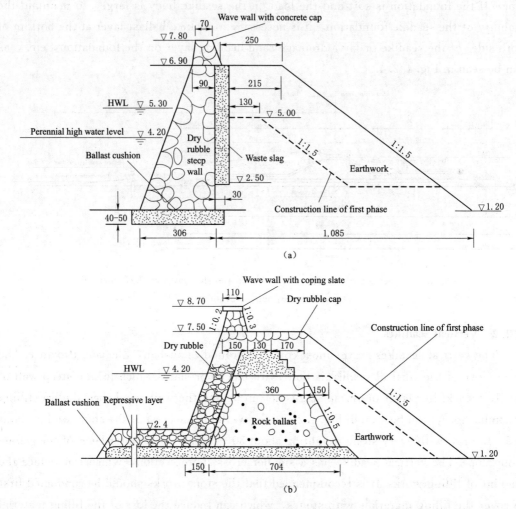

Fig. 4.2.5 Cross-shore section of the vertical seadike
(unit in scale: cm, elevation: m)

4.2.3 Composite seadikes

The water-facing side of this type of seawall is composed of a combination of a steep wall and a sloping section. There are two different combinations. One with a slope on the upper part of the waterfront and a steep wall on the lower part, with the elevation of the crest of the vertical wall tidal levels, as can be seen in Fig. 4.2.6. The other is a steep wall on the upper part of the waterfront with a sloping rock throwing prism below the steep wall. The elevation of the rock throwing prism should be set as high tidal levels, as can be seen in Fig. 4.2.7.

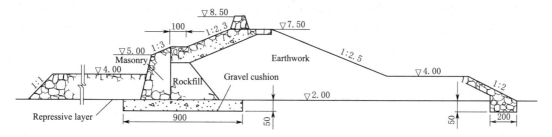

Fig. 4.2.6 Design section of the seawall at Wenling East Sea Pond
(unit in scale: cm, elevation: m)

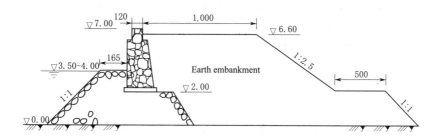

Fig. 4.2.7 Seadike at Sanbaimen, Raoping, Guangdong Province
(unit in scale: cm, elevation: m)

Composite seadikes combine the characteristics of both sloping and vertical seadikes. If the upper and lower slopes are properly combined and applied, the advantages of both sloping seadikes and vertical seadikes can be exploited.

Various factors should be considered when determining the type of seadikes, including topography, geology, hydrology, materials source and the conditions of construction, etc. After comparing every scheme, the most reasonable scheme will be chosen which suits the geographical conditions and can make the best use of seawalls. In general, vertical seadikes are chosen when dike lines are located outside the surf zone or at shallow water depth where the seadikes are affected by sanding waves mostly. If dike lines are located near the surf zone where actions are intense, sloping seadikes will be more appropriate.

Wave dynamics are important factors when choosing the type of seadikes. Physical models are necessary tobe conducted by comprehensively considering the wave shapes, wave runups, wave forces, wave kinematics and dynamics before the selection of seadikes.

4.3 Basic size of the seadike section

4.3.1 Elevation of seadike crest

The elevation of the seadike crest is directly related to the safety of the seadike, the size of the engineering and the investment. Hence, it is an important criterion. The elevation of the seadike represents the elevation of the dike crest after it has sunk and stabilized. For seadikes with breakwaters, the elevation of the seadike is the elevation of the breakwater.

There are many factors that determine the elevation of the dike crest. The most important factors among them are return periods of waves and tides as well as the cumulative frequency of wave trains which are reflected in the seadike defense standards. These factors represent the importance and the extension of the protection of the seadike, as well as the different responses of various seadikes to tidal and wave actions. Other factors such as local meteorological and hydrological conditions as well as the specific structural characteristics of the seadikes can affect the wave characteristics and runup in front of the seadike, and thus affect the determination of the dike crest.

The elevation of the dike crest can be determined as:
$$Z_p = h_p + R_u + \Delta h \tag{4.3.1}$$
Where, Z_p is the crest elevation with design frequency of P, m; h_p is the design high tidal level with frequency of P, m; R_u is the wave runup with design waves, m; Δh is the safe overtopping value, m. Values of Δh are adopted according to relevant standards. $\Delta h = 0$ is adopted when part overtopping is allowed, while $\Delta h = 0.5 - 0.7$ m is adopted when no overtopping is allowed.

On designing the cumulative frequency of wave runup, it is recommended that 2% is used for no overtopping and 13% is used for partly overtopping when irregular wave runup caluculation method is adopted. Wave height with different cumulative frequency $H_{F\%}$ is set as the calculation wave height when regular wave runup caluculation method is adopted. $H_{F\%}$ is set as $H_{2\%}$ when no overtopping is allowed, while $H_{F\%}$ is set as $H_{13\%}$ as overtopping is partly allowed.

With the development of economy in our country, the importance of many areas protected by seadikes along the coast is increasing. It is worth investigating that the design wave cumulative frequency and safe overtopping value for calculating the wave runup that frequently used in the past can be adjusted accordingly. For example, the safe overtopping values in some references suggest $\Delta h = 0.3 - 0.5$ m when overtopping is partly allowed, and $\Delta h = 0.5 - 0.1$ m when overtopping is not allowed. These suggested values are all larger than the dike standard from Zhejiang Province and Fujian Province.

When designing the seadike projects, sometimes the historical high tidal level is used as the design high tidal level rather than frequency analysis due to the lack of the hydrology information. For example, the aforementioned *Technical Regulations for the Design of Reclamation Projects in Fujian Province* provides that the elevation of the dike crest should be determined by comparing these two methods, which adopt the historical high tidal levels and the frequency analysis method to determine the design tidal level, respectively, as can be found in Table 2.1.5 and Table 2.1.6.

4.3.2 Width of dike crest

The width of the dike crest (excluding the breakwater) depends mainly on the stability of the seadike, the stability of the foundations, the impermeability requirements, the construction and flood control requirements and the traffic requirements. The width of the dike crest is generally no less than 3 - 4 m, and it increases to 4 - 6 m for the large reclamation projects. On the mudflat of the Jiangsu coast, the width of the dike crest is generally 6 - 8 m, as the mud on which the dike is built is easily eroded by wind waves. Seadikes of important industrial companies, such as the Shanghai Petrochemical General Factory Seadike, has a crest width of about 10 m. If the crest is combined with road traffic, it should be determined according to the traffic requirements, generally the dike crest road with two-way traffic, the width of the dike crest can be set at 8 m. The section of seadike from Qijiadun to Jinshanzui in Hangzhou Wan, which is combined with the Shanghai-Hangzhou Highway, has a crest width of 20 m or more.

4.3.3 Side slope of seadike

The side slope of seadike is mainly determined by the stability requirements. The section type, armour type and material, embankment material, wave action, foundation conditions and construction conditions are also considered.

The slope can be initially determined according to Table 4.3.1, followed by stability and wave runup calculations to determine a reasonable seadike slope.

Table 4.3.1 Inner and outer slopes of the seadike

Type of armour		Gradient of outer slope	Gradient of inner slope
Sloping seadike	Armour layer with dry blocks	1 : 2.0 - 1 : 3.0	Above water: 1 : 1.5 - 1 : 3.0 Underwater: Sand-mud mixture 1 : 5 - 1 : 10 Mountain clay 1 : 5 - 1 : 7
	Armour layer with slurry blocks	1 : 2.0 - 1 : 2.5	
	Armour layer with ripraps	No less than 1 : 1.5	
	Armour layer with artificial blocks	1 : 1.25 - 1 : 2.00	
Vertical seadike (breakwater)		1 : 0.3 - 1 : 0.5	
Composite seadike		According to the blocks and vertical wall	

For dike without the armour layer, its outer slope is mild. The slope of the clay dike is 1 : 3 - 1 : 2. The slope of clay dike in north Jiangsu is 1 : 15 - 1 : 5. Especially, the outer slope of the seadike in Rudong, Jiangsu is 1 : 25.

When the outer slope is 1 : 1.5 - 1 : 2.0, wave runup can be probably large in the range of wave steepness for common wind waves. However, the engineering work would be increased with the outer slope decreases. Therefore, for larger projects, a reasonable slope should be determined through economic and technical comparison.

In order to reduce wave runup, a wave damping platform can be set on the outer slope. The elevation of the platform should be set at the high tidal level, at which the wave damping effect is better. The width of the platform is usually set as 1 - 2 times of the design wave height, and no less than 2 - 3 m, which is more effective in dissipating waves. However, the width of the platform will also increase the volume of work, so the platform elevation and width of large projects are best determined by experimental comparison.

When the dike is built on soft ground, if the dike is high, it is also necessary to set platforms on both sides of the seadike to increase stability. The platform can be considered in combination with traffic, maintenance and flood control requirements. A platform can be set up every certain elevation, e.g., 5 m, with the platform width larger than 1.5 m.

4.4 Construction of seadike

4.4.1 Dike crest and breakwater

If the dike crest is combined with the highway surface, it should be designed as a road structure according to the requirements of the traffic department. If it is not combined with the highway, protective measures should also be taken to prevent erosion by rainfall and overtopping waves. Generally, dike crest can be protected by mortar, rubble armour, and some important dike crest can use concrete slabs as the armour. For large and medium-sized engineering, it is appropriate to set up a shoulder stone in the inner side of the dike crest, make the length direction of the slates perpendicular to the axis of the seadike, and implement a layer of slates along the inner edge of the dike crest. The armour of seadike should tilt to the inner side with a slope of 2%-4%. And a vertical and horizontal drainage system should be set on the inner slope and the platform of inner slope which is used for draining rain and seawater away.

The breakwater is usually set outside of the dike crest, constructed with dry or slurry blocks and slates, as well as the prefabricated concrete blocks capped by fine stone concrete. For the section of seadike at which wind waves are not strong or the subsidence is large, dry blocks are ought to be used for construction. Especially after the completion of the seadike built on the soft foundation, dry blocks can be installed as the breakwater first. Then the section of the breakwater confronting large waves can be rebuilt by slurry blocks when subsidence of the seadike tends to be stable after a few years.

The height of the breakwater is generally 0.8 - 1.2 m, with the bottom width of 0.8 - 1.2 m, and the top width of 0.6 - 1.0 m. The waterfront surface of the breakwater is a vertical wall and the reverse side is a steep wall with a slope of 1 : 0.2 - 1 : 0.5 usually. The wave wall should have an embedment depth which is not less than 0.3 m below the dike crest and be connected to the seadike as a whole to improve its stability. A common structure type of breakwater can be seen in Fig. 4.4.1.

4.4 Construction of seadike

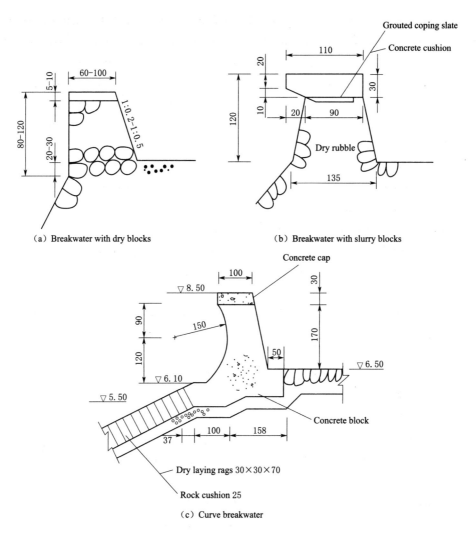

Fig. 4.4.1 Types of crest of the breakwater (unit in elevation: m, scale: cm)

A curved waterfront slope can pick up waves effectively and reduce the amount of wave overtopping significantly, as can be seen in Fig. 4.4.1 (c). Under wind actions, the picked-up water can also likely be blown into the seadike. The main problem of curved wall is that the vertical component of the wave stress increases which is not conducive to its stability. The curved shape also brings certain difficulties to the construction.

After the typhoon No. 13 in 1974, it is found that the elevation of part of the seadikes in Shanghai and Zhejiang Province was not enough. However, due to various limitations, there are difficulties in further raising the dike crest. To achieve this, the wave barrier wall is relocated from the wave-facing side to the inner side, and the top of the embankment is utilized as a wave-dissipation platform. This allows for enhanced resistance to wind and waves without increasing the elevation of the embankment crest.

In wave damping seadike project in reclamation area of Xiaoshan, Zhejiang Province, plain concrete empty boxes of 0.5 m width, 0.8 m height and 1.0 m length are used as the rear-side wave mitigation wall. While the dike crest of 8 m width made of concrete blocks of 1.0 m×1.0 m×0.1 m was used as the wave damping platform. This backward breakwater arrangement needs to strengthen the wave resistance of the top structure of the dike, and also has a greater impact on the crosswalk. Consider locating the road on the inner slope of the earthen embankment, which is more economical than constructing a separate road.

4.4.2 Armour

The main functions of the armour is to protect the clay dike from erosion induced by wind waves, tidal currents and rainfalls.

The basic requirements for the armour layer include:

(1) Under the action of waves, currents and infiltration pressure, the armour structure can maintain stability without damage, so the armour should have sufficient thickness and weight.

(2) Under the action of infiltration in the seadike, the soil of the dike body will not be lost from the gap of the armour blocks, so a backfilter layer or transition layer should be installed under the armour.

(3) Adequate protection should be provided to avoid endangering the safety of the dike after the toe or top of the slope is scoured.

(4) The building materials should be collected as local as possible. Simple construction, easy maintenance and economic cost are recommended.

There are several types of armour, e.g., block armour including riprap, dry block, and slurry block, precast concrete slabs or cast-in-place concrete armour, asphalt concrete armour, artificial block armour, cement soil and turf armour.

At present, rubble slopes are the most commonly used in sea dike revetments, especially masonry slopes. Masonry slopes can be divided into two types of structures based on whether cement mortar is used for construction: mortared rubble slopes and dry-laid rubble slopes, each having its own characteristics. Generally, dry block armour protection is more flexible. It can adapt to the deformation of the seadike, and can be found in time when the seadike soil loss generated by local evacuation. However, its integrity is poor, the ability to resist wind, waves and tides is relatively weak, and the dike body soil is easily lost under the action of infiltrating water and waves. In contrast, slurry block armours have better integrity and are more resistant to waves and currents. However, the flexibility is poor, and local subsidence cause the collapse and damage of the armour. At the same time, the porosity of the slurry armour decreases, resulting in an increase in wave runup on the slope.

The thickness of the block slope and the weight of individual block should be determined by calculation (see "Seadike Design"), but its minimum thickness should not be less than 30 – 40 cm. The slurry block armour should set deformation joints and drainage holes, the longitudinal moment of deformation joints is 10 – 15 m, and the longitudinal and horizontal spacing of drainage holes is 2 – 3 m. The slurry block armour is generally built with No. 10 cement mortar and No. 15 cement mortar, and the width of block joints is 2 – 3 cm. Concrete-filled blocks are also used to protect the slope by pouring 100-grain concrete in between the blocks. The block spacing is required to be not less than 2 – 3 times the maximum diameter of concrete aggregate, this kind of armour is easy to construct.

Mechanical pounding can be used to improve the quality of the engineering, the artificial roughness of slope can also be used to dissipate waves. The bar stone can be built from blocks to enhance the ability of mitigating waves for the blocks due to the limitation on the weight of manually moving blocks (200 – 250 kg). The thickness of the bar stone can be up to 0.6 – 2 m, which can mitigate waves with height up to 3 m.

Stone armour protection is to fill the slope surface with appropriate weight graded blocks, without artificial masonry. Sometimes only minor adjustments are made to increase the stability and interlocking conditions of the blocks. The stone throwing armour has strong ability to adapt to the deformation of the dike body and good wave dissipation performance. However, the overall resistance of the armour to wind waves is poor because the embedded effect between the blocks is poor. Individual block should not be too heavy and too large, so it can generally be used as temporary protection. If used as a permanent armour, it should only be used in places where the wind waves are not too strong or in toe protection projects. The thickness of the stone layer is generally 50 – 90 cm.

Concrete slab armours can be divided into two types: precast and cast-in-place. The shape of precast concrete slab is generally square or rectangular, with size of 0.5 m×0.5 m or 1.0 m×1.0 m, thickness of 15 – 20 cm. Generally, they can be made into the enterprise mouth seam. The seam form has two types, i. e., lap type and tongue-groove type. The experiments conducted by USA. Coastal Engineering Center showed that the lap type is more stable than the tongue-groove type. Integral cast-in-place concrete slab has size from 5 m×5 m to 20 m×20 m, with thickness of 15 – 40 cm. The joints of the slab are filled with asphalt concrete or profiled rubber. The armour has a certain degree of flexibility, so that when the sink deformation of the seadike occurs, it can still maintain its function. The integral concrete armour is either cast directly on the clay slope or based on a whole bedding layer, depending on the characteristic of the clay and construction conditions.

Artificial block armours are used in more important seadikes such as four-legged tapered armour in Shanghai Petrochemical General Factory Dike, Four-legged hollow squares, fence panels used in Shanghai Baoshan Iron and Steel General Factory Riverdike.

Cement soil armour was used in the early days of highway construction, and after the

1950s it was gradually used in the United States for earthen dams to prevent waves and arch slopes, and it is an economic armour type in areas lacking stone. Most soils can be used to make cements, among which the sediments with gravels are most useful. The cement would increase when there is a lack of fine particles of soils on the beach. Typical cement content is 10%–12% of the compacted volume of the soil, but it can vary between 7% and 15% depending on the soil quality. Cement armour requires less sinkage of the dike body. The cement layer is composed of several horizontal layers by pouring and compaction. The thickness of each layer after compaction is 15–20 m, the general total thickness is 0.6–0.9 m, the combination between layer and layer must be very well-combined, otherwise it is easy to damage by waves.

4.4.3 Armour layer

In order to prevent the loss of dike soil under the action of waves and infiltration, an armour layer or transition layer should be set between the surface protection blocks and the soil as the foundation of the armour. The design of the armour layer can follow the design of the inverted layer. When the design requirements are difficult to implement in construction, the filter layer can be used as an inverted layer with gravels or rock ballasts. However, it is better to set an inverted layer among the blocks, sediments and clays.

The normal inverted layer is composed of two or three layers of non-cohesive materials. Their particle size increases layer by layer along the infiltration direction (from inside to outside the dike). If gravel with a particle size of 3–5 cm is used as the material for the filter layer of the clay dike, its thickness should not be less than 20–30 cm. When using natural graded stone slag as a transition layer, the long side of the stone slag should be less than 15 cm, and the thickness of the stone slag layer is not less than 50 cm. If the dike body is filled with sandy soil with high sand content, the thickness of the filter layer should also be increased to 60–80 cm. For example, the thickness of the bedding layer of reclamation in Qingtian Gang, Fujian Province is 60 cm, and gravel with a diameter of less than 7 cm is used as the materials of the armour layer.

When the design of the armour layer follows the design of the inverted layer, the materials at the first layer can be determined by the following equation:

$$\frac{D_{15}}{d_{85}} \leqslant 4-5, \quad \frac{D_{15}}{d_{15}} \geqslant 5 \tag{4.4.1}$$

Where, D_{15} is the grain size of the materials of the inverted layer, the subscript 15 means that 15% of the total weight of particles smaller than this size; d_{15}, d_{85} are grain sizes of protected clays, the subscript 15 and 85 mean 15% and 85% of the total weight of particles smaller than that particle size.

When determining the second and third inverted layers, the above method is still followed. For the second layer of inverted layer material, the first layer of inverted layer material is used as the protected soil, and so on.

Some design manuals in the United States stipulate that the inverted layer set up between the stone throwing armour and the dike body clay is with $D_{85} \geqslant 2.5 - 3.8$ cm when the maximum wave height is less than 1.2 m. When the wave height is 1.2 - 3.0 m. D_{85} is noted between 3.8 cm and 5.0 cm. If the dike material already meets the above requirements for D_{85}, no additional inverted layer is required.

In addition, the minimum thickness t is set for a single inverted layer. When the maximum wave height is less than 1.2 m, t is set as 15 cm. When the maximum wave height is between 1.2 m and 2.4 m, t is adopted as 23 cm. When the maximum wave height is between 2.4 m and 3.7 m, t is set as 30 cm.

Some calculation methods of the Soviet Union provided the standard for the unevenness coefficient D_{60}/D_{10} of armour layer materials, generally set as 5 - 12. It can be set as 5 - 8 when the wave height is larger than 2 m. In addition, the interlayer coefficient D_{50}/d_{50} for each layer is also specified. For the layer where the armour layer is in contact with the soil slope, $D_{60}/D_{10} \leqslant 10$.

The above method of determining the armour layer material is mainly applicable to non-cohesive soil. There are some different specifications for cohesive soil. For example, according to the relevant norms of the former Soviet Union, if the clay particles ($\leqslant 0.005$ mm) containing more than 30%, plasticity index $\geqslant 7$, water content greater than the plastic limit of clayey soil, the first layer of the single layer of mat or multi-layer mat (i.e., a layer in direct contact with the soil slope), the uneven coefficient of the filter layer material $D_{60}/D_{10} \leqslant 20$, and the particle composition of the filter layer material is no longer determined by the interlayer coefficient. Thus, it is stipulated that: when the clay is not allowed to be eroded, the particle size of the filling particles $\leqslant 2 - 3$ mm. If the clay is allowed to be slightly eroded at the contact with the armour layer, the particle size of the filler material is $\leqslant 6 - 8$ mm, etc. Except for the first layer, the design method for each layer of the armour layer and the thickness of the first layer is the same as that for the armour layer of non-cohesive soil slopes.

In the late 1950s, geotextiles with synthetic fibers or plastic meshes began to be used in coastal projects. They have been widely used as a filter layer for seawalls and dikes in the 1970s. Geotextiles are generally laid close to the soil material. This can reduce the thickness of the armour layer, eliminating the gravel transition layer of about 20 - 30 cm thick. Then lay sand and gravel material between the armour layer and the geotextile for the transition layer, otherwise they may be torn if large stones are placed directly on the geotextile. However, if the content of silt or clay in the soil exceeds 50%, it is possible for fine particles to cross or block the mesh under the action of fluctuating water flow, which increases the hydrostatic pressure in the dike. At that time, a coarse sand cover layer can be added between the soil slope and geotextile, the thickness of which is about 15 – 20 cm. Generally, the total cost of thin sand bedding layer combined with geotextile slope protection is more economical than using multi-layer sand and gravel bedding layer without geotextile, and it is easy to control the construction quality.

Using geotextiles for the armour layer, the corresponding geotextiles should be selected according to the particle size distribution of the protected soil material to meet the requirements of water permeability and soil retention. The stability of the fabric sliding along the soil slope should be considered.

4.4.4 Footings of armour layer

To ensure the stability of the seadike, footings should be set at the lower end of the seadike. The role of the footing is mainly to support the dike and prevent it from slipping along the main slope surface. It also protects the toe of the dike from the strong scouring that may occur under wave action. The structure types of footing include the buried footing commonly known as large square foot, as can be seen in Fig. 4.4.2 (a) and (b) or stone throwing prisms in Fig. 4.4.2 (c) and pile stone footing [Fig. 4.4.2 (d)], etc.

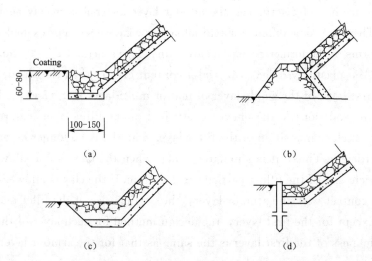

Fig. 4.4.2 Structural types of footing of armour layer (unit: cm)

The buried footing is suitable for mudflat with high elevation, and the stone throwing prisms are used to low level flats. The depth of the footing foundation should be determined according to the topography of the beach and local wave conditions, etc. The depth of the bottom-flow scouring trough in front of the foot of the Shanghai seadike is generally a few tens of centimeters, so the depth of the berm foundation can generally be considered as 1 m below the beach surface. The estimation of the maximum scour depth that may be caused in front of a building under wave action is a complex and difficult problem to be calculated precisely. Some references (e.g., Shore Protection Manual, CERC) suggest the maximum depth of the scour is approximately equal to the maximum wave height that can be supported by the original water depth at the foot of the dike, i. e., local critical breaking wave height.

If the mudflat is low and it is difficult to build the foundation to the required depth, it is also possible to adopt the way of protecting the beach with firewood row of pressed stones before the foot of the slope, so as to raise the foundation of the slope appropriately. In the section of strong wind waves, rock throwing, piling and other foot protection procedures can be used at the same time. For example, the Jinshanwei seadike is protected by reinforced concrete sheet piles at the foot of the dike and by ripraps on the outside of the sheet piles, with a thickness of 1 m and a top width of 3.5 m, as shown in Fig. 4.4.3.

The old sea ponds of the Qiantang River commonly use anti-wash protection tans (locally called Tanshui) to protect the footings of the ponds. The construction of the protection tans can generally only be carried out above the low tidal level. When the bottom in front of the dike is further eroded deeper, there is a risk that the timber piles will topple over and cause damage to the protection tans.

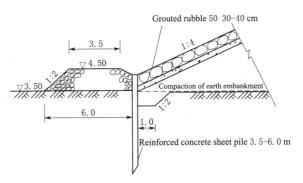

Fig. 4.4.3 Footing of armour layer at Jinshan Weihai (unit: m)

Therefore, in addition to the original protection tank and a lower protection tank should be built, called the second tank. The old sea ponds of the Qiantang River commonly have two and even three tanks.

The footing project is very similar to the armour project in the shore protection, so there is a detailed introduction in the Chapter 2.

4.4.5 Protective wall

The protective wall is the main structure of a vertical seadike, usually a gravity retaining wall. As part of the seawall structure, it mainly withstands waves and currents and protects the dike filling behind the wall, so it plays the role of seadike surface protection. At the same time, it supports the soil pressure from the dike body to maintain the stability of the dike soil.

The outer slope of protective wall of the vertical seadike is generally 1 : 0.5 − 1 : 0.3. As shown in Fig. 4.4.4, the interior of the protective wall can be built with dry block, the sea front should be used in the masonry bars, the bars will be built into a stepped shape. The size of the layer of masonry bars is generally 20 cm × 25 cm × 60 cm − 30 cm × 30 cm × 80 cm. The bottom width of the protective wall generally shall not be less than 1/2 wall height, with width of 0.8 – 1.2 cm. Protective wall of the top 30 cm × 30 cm × 140 cm stone vertical long 450 wooden pile two rows of low tidal level, 30 cm × 30 cm × 140 cm stone flat gravel bedding thick 20 part should be considered with the wave wall connected into a whole, the wall should be set up under the gravel bedding, beach surface should also be set up when the low bed of piled stone. Gravel or slag transition layer with thickness of generally 0.6 – 1.0 m should be set up between the back of the protective wall and the fill. If dry block is used as the protective wall surface layer, a heavy weight block should be selected to enhance its stability against waves. There is a possibility that dry block or stone block will be pulled out by the negative wave pressure in the section of dike with large wind waves, which can be improved by cement grouting of dry block or stone block joints.

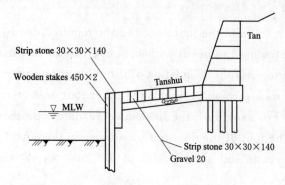

Fig. 4.4.4 Anti-erosion Tanshui at Qiantang River pondm (unit: cm)

4.5 Design of seadike

4.5.1 Calculation of wave runup

The vertical height of the wave climbing along the slope of the building (calculated from the still tidal level) is called the wave runup. There are many ways to calculate the wave runup, which are described as follows.

1. Method of breakwater and revetment design specifications

This method is applicable to waves acting positively on the slope of the seadike, with slope $m = 1 - 5$, water depth $d = (1.5 - 5.0)H$ in front of the building, water bottom slope i is less than or equal to 1/25. Wave runup for regular waves are calculated as follows.

$$R = K_\Delta K_d R_0 H \tag{4.5.1}$$

Where, R is the wave runup, m, starting from the still tidal level; K_d is the water depth correction factor, as provided in Table 4.5.1; K_Δ is the coefficient of roughness permeability related to the armour structural type, as provided in Table 4.5.2; R_0 is the wave runup when $K_\Delta = 1$ and $H = 1$ m, it is related to the slope $m = \cot\alpha$ and wave steepness L/H, as provided in Fig. 4.5.1; H is the progressive wave height at the seadike, m.

4.5 Design of seadike

Table 4.5.1   Water depth correction factor

$\dfrac{d}{H}$	1.5	2.0	3.0	4.0	5.0
K_d	1.12	1.15	1.00	0.96	0.94

Table 4.5.2   Coefficient of roughness permeability

Armour structural type	K_Δ
Smooth impermeable surface protection (asphalt concrete)	1.00
Concrete armour	0.90
Masonry	0.75 – 0.80
Blocks (1 layer)	0.60 – 0.65
Four-legged hollow cube (1 layer)	0.55
Blocks (2 layers)	0.50 – 0.55
Concrete blocks (2 blocks)	0.50
Four-legged cone (2 blocks)	0.40
I-shape block (2 blocks)	0.38

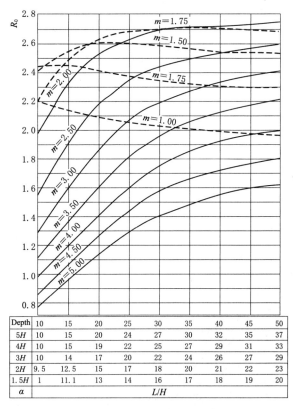

Fig. 4.5.1   Curves for determining wave runup

2. Method of the Former Soviet Union Standards (СНиП Ⅱ 2.06.04 – 82 ∗)

For the positive incoming wave, the water depth d in front of the slope satisfies the condition of $d \geqslant 2H_{1\%}$, the wave runup on the slope with a cumulative frequency of 1‰

Chapter 4 Seadike Engineering

should be determined by the following equation.

$$R_{1\%} = K_1 K_2 K_3 K_4 H_{1\%} \tag{4.5.2}$$

Where, $R_{1\%}$ is the wave runup with cumulative frequency of 1%, m; K_1 and K_2 are the roughness and permeability of the slope, as provided in Table 4.5.3; K_3 is the wind coefficient, as provided in Table 4.5.4; K_4 is the wave steepness and slope coefficient, as provided in Fig. 4.5.2; $H_{1\%}$ is the wave height with cumulative frequency of 1%, m; $(H_0)_{1\%}$ is the deep-water wave height with cumulative frequency of 1%, m.

Table 4.5.3   Roughness and permeability of the slope

Armour structure	Relative roughness $r/H_{1\%}$	Coefficient K_1	Coefficient K_2
Concrete (reinforced concrete) slabs Gravel, pebbles, blocks of stone or concrete (reinforced concrete) block armour	—	1.00	0.90
	<0.002	1.00	0.90
	0.005 – 0.010	0.95	0.85
	0.020	0.90	0.80
	0.050	0.80	0.70
	0.100	0.75	0.60
	>0.200	0.70	0.50

Note: Roughness characteristic scale $r(m)$ should be used equal to the average particle size of armour material or the average scale of concrete (reinforced concrete) block.

Table 4.5.4   Wind coefficient

	$\cot\varphi$		1 – 2	3 – 5	>5
K_3 Coefficient	Wind speed V_ω /(m/s)	≥20	1.4	1.5	1.6
		10	1.1	1.1	1.2
		≤5	1.0	0.8	0.6

Note: φ is the inclination of the slope to the horizontal (degrees).

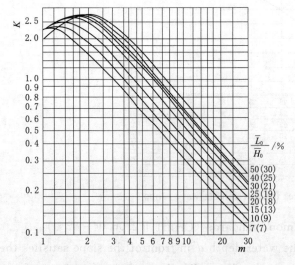

Fig. 4.5.2 Curves of coefficient K

In above figures and tables $m = \cot\alpha$, α is the angle of the slope, L_0 is the deep-water wave length. When the water depth in front of the dike $d < 2H_{1\%}$, the wave steepness determined in brackets in Fig. 4.5.2 and under the condition of water depth $d = 2H_{1\%}$ must be used to determine the value of K. The r in Table 4.5.3 is the roughness, which equals the average particle size of the armour material or the average size of the concrete and reinforced concrete blocks.

4.5 Design of seadike

The cumulative frequency of waves on the slope is the value of i of the wave runup, which must be determined by multiplying the value of $R_{1\%}$ obtained from Eq. (4.5.2) by the coefficient K_p listed in Table 4.5.5.

Table 4.5.5　　　　　　　　　　Values of coefficient K_p

Cumulative frequency of wave runup $i/\%$	0.1	1.0	2.0	5.0	10.0	30.0	50.0
Coefficient K_p	1.10	1.00	0.96	0.91	0.86	0.76	0.68

When the wave propagates toward the building with β angle (β is the angle between the wave direction line and the normal of the slope), the wave runup should be multiplied by the factor K_β, see Table 4.5.6.

Table 4.5.6　　　　　　　　　　Values of coefficient K_β

$\beta/(°)$	0	10	20	30	40	50	60
K_β	1.00	0.98	0.96	0.92	0.87	0.82	0.76

3. Putian method

Under the direct action of wind, the wave runup of a positive incoming wave on a single slope of $m = 1.5 - 5.5$ can be determined by the following equation.

$$R_F = \frac{K_\Delta K_V K_F}{\sqrt{1+m}} \sqrt{\overline{H}\,\overline{L}} \qquad (4.5.3)$$

Where, R_F is the wave runup with cumulative frequency of F, m; K_Δ is the coefficient of roughness permeability related to the armour structural type, as provided in Table 4.5.7; K_V is determined by the dimensionless parameter combined with wind speed V, water depth d in front of dikes, and gravity acceleration g, as provided in Table 4.5.8; K_F is the wave runup cumulative frequency conversion factor, as provided in Table 4.5.9; $\overline{H}$ and $\overline{L}$ are mean wave height and wave length in front of the dike, m.

Table 4.5.7　　　　　　　　　　Coefficient of roughness permeability K_Δ

Armour type	K_Δ
m	1.00
Concrete and concrete slab armour	0.90
Turf armour	0.85 - 0.90
Masonry armour	0.75 - 0.80
2 layers of throwing blocks (impermeable foundation)	0.60 - 0.65
2 layers of throwing blocks (permeable foundation)	0.50 - 0.55

Table 4.5.8　　　　　　　　　　Values of K_V

$\frac{V}{\sqrt{gd}}$	≤1.0	1.5	2.0	2.5	3.0	3.5	4.0	≥5.0
K_V	1.00	1.02	1.08	1.16	1.22	1.25	1.28	1.30

Table 4.5.9　　　　Wave runup cumulative frequency conversion factor

$\dfrac{\overline{H}}{d}$	$F/\%$	1	2	3	4	5	10	13	20	30	50
<0.1	K_F	2.23	2.07	1.97	1.90	1.84	1.64	1.54	1.39	1.22	0.96
0.1–0.3		2.08	1.94	1.86	1.80	1.75	1.57	1.48	1.36	1.21	0.97
>0.3		1.86	1.76	1.70	1.65	1.61	1.48	1.40	1.31	1.19	0.99

4. Normative revision law for harbor engineering

As the name implies, this calculation method is a revision of the hydrographic code for ports and waterways regarding the calculation method of wave runup. Its features are mainly in three aspects:

(1) The revision is based on the results and experiences of many years of research on wave action on slope dike at Hohai University, and the research involves a greater range of variation in the various influencing factors such as slope $m = 0 - 20$, wave steepness $\dfrac{L}{H} = 10 - 50$, relative water depth $\dfrac{d}{H} = 2.5 - 25$. Thus, the range of application is also wider.

(2) The revised provisions, in addition to giving the calculation method of wave runup, also give the calculation method of falling depth and wave rundown amplitude. It also changes a large number of graphs and tables in the original method into functional equations for computer applications.

(3) The originally established relationship between wave runup and the variation of each single factor has been improved to a composite factor composed of these single factors, i.e., the Irribaram number $I_r = \dfrac{\tan\alpha}{\sqrt{\dfrac{H}{L_0}}}$. It is shown that the I_r number is a very effective physical parameter to reflect a series of motion and dynamic characteristics of waves (e.g. reflection, runup, pressure, etc.) during the interaction between waves and sloping structures.

The method of calculating the wave runup by the code revision method is as follows.

The tidal level change R of an incident regular wave on a slope, including the wave runup R_u and wave rundown R_d, can generally be determined by Eq. (4.5.4).

$$R = K_\Delta R_1 H \quad (4.5.4)$$

Where, K_Δ is the coefficient of roughness permeability, as provided in Table 4.5.2; R_1 is the wave runup R_{u1} or wave rundown R_{d1} when $H = 1$ m. It is related to sloping factor M and can be determined as follows.

$$R_1 = K_1 \tanh(0.432M) + [(R_1)_m - K_2] R(M) \quad (4.5.5)$$

$$M = \dfrac{1}{m}\left(\dfrac{L}{H}\right)^{\frac{1}{2}} \tanh\dfrac{2\pi}{L}d^{-\frac{1}{2}} \quad (4.5.6)$$

4.5 Design of seadike

$$(R_1)_m = \frac{k_3}{2} \tanh \frac{2\pi d}{L} \left(1 + \frac{\frac{4\pi d}{L}}{\sin \frac{4\pi d}{L}}\right) \quad (4.5.7)$$

Where, $(R_1)_m$ is the maximum wave runup or rundown for a certain $\frac{d}{L}$.

Three constants $K_1 - K_3$ can be determined from Table 4.5.10.

Table 4.5.10 K_1, K_2 and K_3

Wave phase	K_1	K_2	K_3
Runup	1.240	1.029	4.980
Rundown	−0.760	−0.735	−2.760

When calculating wave runup, $R_u(M)$ is used for $R(M)$ in Eq. (4.5.8).

$$R_u(M) = 0.350 M^{1.954} \exp(-0.42M) - 7.801 M^{2.02} \exp(-2.96M) \quad (4.5.8)$$

When calculating the wave rundown, $R(m)$ is a function of wave rundown.

$$R_u(M) = 0.350 M^{1.954} \exp(-0.42M) - 7.80 M^{2.02} \exp(-2.69M) \quad (4.5.9)$$

5. Wave runup on composite sections

For the composite slope section with platform, the converted slope m of the section can be determined first, and then its wave runup can be approximated by the single-slope section with slope m_x.

As can be seen in Fig. 4.5.3, the slopes above and below the platform are noted as m_{up} and m_{down}. The water depth over the platform is d_w, which is negative if the still tidal level is below the platform. B is the width of the platform and $\overline{L}$ is the mean wave length.

If $\Delta m = m_{down} - m_h = 0$, i.e., consistent slope above and below the platform, then:

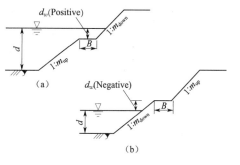

Fig. 4.5.3 Composite slopes with platform with d_w (upper panel) and negative d_w (lower panel)

$$m_x = m_{up}\left(1 - 4.0 \frac{|d_w|}{L}\right) K_b \quad (4.5.10)$$

if $\Delta m > 0$, then:

$$m_x = (m_{up} + 0.3\Delta m - 0.1\Delta m^2)\left(1 - 4.5 \frac{d_w}{L}\right) K_b \quad (4.5.11)$$

if $\Delta m < 0$, then:

$$m_x = (m_{up} + 0.5\Delta m + 0.08\Delta m^2)\left(1 + 3.0 \frac{d_w}{L}\right) K_b \quad (4.5.12)$$

K_b in Eq. (4.5.11) to Eq. (4.5.12) can be determined as follows:

$$K_b = 1 + 3 \frac{B}{L} \quad (4.5.13)$$

The above calculation method for wave runup of compound slope section is applicable under the following conditions.

Generally, the location of the platform or up and down slope turning point should be set near the still water level or slightly above the still tidal level, where the platform energy dissipation effect is best. Hunt suggested that the width should be no less than $\frac{1}{5}$ of wave length, i.e., $\frac{B}{L} \geqslant \frac{1}{5}$ for regular waves. Delft Laboratories recommended that the wave runup with a platform should be equal to the wave runup without a platform multiplied by a factor $1-\frac{B}{L}$.

Under the conditions of $0 < \Delta m < 1$, $0 < \frac{B}{H} < 7$ and $0 < \frac{d_w}{L} < 0.5$, the converted slope m_x can also be calculated as follows. This method is also called the Peshkin method.

$$\frac{1}{m_x} = \frac{1-0.2\sqrt{\frac{B}{H}}}{m_{down}} + 2\frac{d_w}{L}\left(\frac{1}{m_{up}} - \frac{1-0.2\sqrt{\frac{B}{H}}}{m_{down}}\right) \quad (4.5.14)$$

The Peshkin method is only applicable to sections with little changes in slope, and the lower slope is gentle as well as the upper slope is steep. When d_w tends to 0, m_x is only related to m_{down} and regardless of m_{up} according to Eq. (4.5.14), which is inconsistent with some physical experiments. Studies show that the wave runup is mainly controlled by the upper slope rather than the lower slope when the platform is near the still tidal level.

6. Calculation of the climb when there is ballast in front of the dike

Sloping dikes with ballast is shown in Fig. 4.5.4. Following methods can be adopted when $m \geqslant 1.0$:

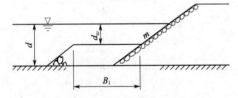

Fig. 4.5.4 Sloping dikes with ballast

(1) Calculate the wave runup without ballast as described above.

(2) The resulting wave runup is multiplied by the ballast correction factor k_y to get the wave runup in the presence of ballast. The values of k_y are shown in Table 4.5.11.

Table 4.5.11　　　　　　　　　　Values of k_y

	B_1/L	0.2			0.4			0.6			0.8		
	$\overline{L}/H$	⩽15	20	25	⩽15	20	25	⩽15	20	25	⩽15	20	25
$\frac{d_w}{H}$	1.0	0.85	0.94	0.99	0.75	0.83	0.87	0.70	0.78	0.81	0.68	0.75	0.79
	1.5	0.92	1.03	1.13	0.86	0.96	1.06	0.81	0.91	1.00	0.79	0.88	0.97
	2.0	0.95	1.10	1.18	0.91	1.06	1.14	0.89	1.01	1.11	0.87	1.01	1.09
	2.5	0.98	1.04	1.10	0.96	1.02	1.08	0.93	0.99	1.04	0.92	0.98	1.03

4.5 Design of seadike

(3) When $\dfrac{d_w}{H} \leqslant 1.5$, $m \leqslant 1.5$, above wave runup is needed to multiplie a factor K_m. The factor can be determined from Table 4.5.12.

Table 4.5.12 Values of K_m

$\dfrac{d_w}{H}$	m	K_m			
		$B_1/\overline{L}=0.2$	$B_1/\overline{L}=0.4$	$B_1/\overline{L}=0.6$	$B_1/\overline{L}=0.8-1.0$
1.0	1.0	1.35	1.26	1.25	1.14
	1.5	1.16	1.10	1.10	1.03
4.5	1.5	1.50	1.60	1.50	1.40
	1.5	1.36	1.46	1.30	1.24

In Table 4.5.12, B_1 is the width of crest of the ballast. The $H_{13\%}$ is used for wave height and $\overline{L}$ is used as wave length for determine the $B_1/\overline{L}$ and d_w/H.

4.5.2 Calculation of armour

Calculation of armour includes: ①thickness of masonry armour; ②weight of a single block of stone throwing; ③slab thickness of concrete armour.

1. Calculation of masonry armour thickness

(1) Port Engineering Technical Specification method. The thickness h(m) of the dry masonry armour of a sloping structure under wave action is generally calculated according to the following formula. For the slurry masonry armour with drainage holes, the same thickness as the dry masonry armour can be used.

$$h = 1.3 K_\gamma H (K_{md} + K_\delta) \dfrac{\sqrt{m^2+1}}{m} \quad (4.5.15)$$

Where, K_{md} is a coefficient related to m and $\dfrac{d}{H}$ of a slope, as can be found in Table 4.5.13; K_δ is wave steepness factor, as can be found in Table 4.5.14.

Table 4.5.13 Values of K_{md}

d/H	K_{md}				
	1.5	2.0	3.0	4.0	5.0
1.5	0.426	0.261	0.130	0.080	0.054
2.0	0.354	0.198	0.087	0.043	0.031
2.5	0.332	0.180	0.076	0.040	0.026
3.0	0.322	0.171	0.070	0.037	0.023
3.5	0.314	0.166	0.067	0.035	0.021
4.0	0.310	0.162	0.065	0.034	0.020

Table 4.5.14 Values of K_δ

L/H	10	15	20	25	30
K_δ	0.081	0.122	0.162	0.202	0.243

Note: For the slurry masonry armour with drainage holes, the same thickness as the dry masonry armour can be used.

$$K_\gamma = \frac{\gamma}{\gamma_b - \gamma} \qquad (4.5.16)$$

Where, K_γ is the gravity effect; γ_b is the heaviness of the block materials, kN/m^3, commonly used material heaviness is shown in Table 4.5.15; γ is the heaviness of water, kN/m^3.

Table 4.5.15　　　　　**Commonly used material heaviness**

No.	Type and name	Heaviness/(kN/m^3)	Note
1	Steel and iron 1) Steel 2) Iron	 78.5 72.5	
2	Concrete 1) Concrete 2) Reinforced concrete	 23.0 – 24.0 24.0 – 25.0	
3	Slurry material stone 1) Granite 2) Limestone 3) Sandstone	 26.0 – 27.0 25.0 24.0	
4	Slurry blocks 1) Granite 2) Limestone 3) Sandstone	 24.0 – 25.0 23.0 – 24.0 22.0	
5	Dry Blocks 1) Granite 2) Limestone 3) Sandstone	 22.0 21.0 20.0	
6	Filling materials 1) Throwing blocks 　　Throwing blocks (underwater) 2) Throwing rubbles 　　Throwing rubbles (underwater) 3) Sand soil (wet) 　　Sand soil (very wet) 4) Fine sand 　　Fine sand (underwater) 　　Medium sand 　　Medium sand (underwater) 　　Coarse sand 　　Coarse sand (underwater) 5) Sand with pebbles (wet) 6) Dust (3:7 or 2:8) 7) Cinder 　　Cinder (underwater) 8) Fly ash 　　Fly ash (tidal range) 　　Fly ash (underwater)	 17.0 – 18.0 10.0 – 11.0 15.0 – 17.0 10.0 – 11.0 18.0 20.0 18.0 9.0 18.0 9.5 18.0 9.5 19.0 18.0 – 19.0 10.0 – 12.0 4.0 – 5.0 12.0 16.0 7.0	 Suitable for particle size less than 0.1 mm Particle content not more than 10% Compacted wet

4.5 Design of seadike

Continued

No.	Type and name	Heaviness/(kN/m³)	Note
7	1) Asphalt gravel pavement 2) Mud-caked gravel pavement 3) Asphalt concrete pavement	18.0 21.0 22.0	Fine sand Medium sand
8	Railroad gravel ballast	20.0	Including upper superstructure

Note: Those not specifically noted in the figure are heaviness above water.

(2) Peshkin method. Under wave action, the slope thickness h (m) of a dry block armour for a sloping seadike of $1.5 \leqslant m \leqslant 5$ can be calculated according to the following formula.

$$h = K K_\gamma \frac{H}{\sqrt{m}} \sqrt{\frac{L}{H}} \qquad (4.5.17)$$

Where, K is a coefficient, which is 0.266 for general dry masonry, and 0.225 for squared stone (including bar); H is the wave height in front of the dike. When $\frac{d}{L} \geqslant 0.125$, $H_{1/4}$ is adopted. When $\frac{d}{L} \geqslant 0.125$, $H_{13\%}$ is adopted, d is the water depth in front of the dike; Other parameters are adopted as mentioned above.

The thickness of slurry block stone can still be calculated by Eq. (4.5.17), when the accumulated frequency of wave height can be taken as $H_{13\%}$.

2. Calculation of weight of a single block of riprap stone

(1) Hudson formula. Hudson formula was proposed by the American Hudson on the basis of a large number of model tests for the stable weight of a single block on the armour at 40 years ago. The formula is provided below:

$$W = 0.1 \gamma_b K_\gamma^3 \frac{1}{K_D} \frac{H^3}{m} \qquad (4.5.18)$$

Where, W is the weight of the single block, t; K_γ is the heaviness factor, as can be seen in Eq. (4.5.16); γ_b is the heaviness, kN/m³; m is the slope, $m = \cot\alpha$; H is the wave height, m; K_D is the stability factor, which is related to armour type, block shape, surface roughness and allowable instability rate, etc. It can be determined by Table 4.5.16.

Table 4.5.16   Allowable instability rate for blocks

Armour block	Structural type	$n/\%$	K_D	Note
Four-legged hollow cube	Set 1 layer	0	14.0	
Block stone	Set 1 (standing) layer	0-1	5.5	
Four-legged cone	Set 2 layers	0-1	8.5	

Chapter 4 Seadike Engineering

Continued

Armour block	Structural type	$n/\%$	K_D	Note
I-block	Set 2 layers	0	18.0	$H \leqslant 7.5$ m
		1	24.0	$H < 7.5$ m
Block stone	Throwing 2 layers	1-2	4.0	
Squared stone	Throwing 2 layers	1-2	5.0	
King block	Throwing 2 layers	0	18.0-24.0	

The allowable instability rate $n(\%)$ indicates the percentage of the number of blocks that are allowed to move and roll off by wave strikes in the range of one wave height above and below the still water surface.

In the case where the blocks are selected and their weight is approximately uniform, the weight of the blocks can be calculated according to Eq. (4.5.18). In addition, this formula is applicable to the case where the overtopping hardly occurs. For the slope building located at the breaking water depth d_h, the weight of the blocks should be 10% to 25% heavier.

If unselected and unsorted graded stone is used as the armour, the weight of the stone is determined by the following formula.

$$W_{50} = 0.1\gamma_b K_\gamma^3 \frac{1}{K_{DD}} \frac{H^3}{m} \qquad (4.5.19)$$

Where, W_{50} is the weight of the blocks, t, 50% of all blocks with less than that weight; K_{DD} is the stability factor of the unorganized blocks, taken as 2.5 when the waves are not broken, and 2.2 when the waves are broken; H is the wave height, taken as the $H_{13\%}$. For unsorted block armour, the design wave height is limited to 1.5 m.

The maximum weight $W_{max}(t)$ and minimum weight $W_{min}(t)$ of graded blocks are:

$$W_{max} = 3.6 W_{50}$$
$$W_{min} = 0.22 W_{50}$$

Hudson formula due to its simple form and being easy to use, is still widely used. China's port engineering specifications also use the formula. The main drawback of Hudson formula is that it fails to take into account the effect of wave period in the formula. The Eq. (4.5.18) can be rewritten as:

$$N_S = (k_D m)^{1/3} = \frac{(0.1\gamma_b)^{1/3} H_{D=0} K_\gamma}{W^{1/3}} \qquad (4.5.20)$$

Where, N_S is the damage factor; $H_{D=0}$ is the critical wave height at which the block does not become unstable under wave actions.

According to Hudson formula, the stable weight of the block is proportional to the third power of the wave height when the block capacity, shape, armour type and slope are determined. Therefore, when a block has a certain weight, its wave height of critical instability is also determined. Reflected in Eq. (4.5.20), N_S should also be a definite value. However, experimental studies have shown that a range of different N_S values can be obtained by simply changing the wave period, with the above factors remained constant. It is well known that different combinations of wave periods with a certain slope m give different numbers of $I_r\left(I_r=\dfrac{1}{m\sqrt{H/L_U}}\right)$. The N_S-I_r relationship curve is a concave curve with the minimum value of N_S occurring roughly between $I_r=2$ and 3. This suggests that although the block weight remains unchanged, as do other slopes, block capacities, shapes, etc., the change in wave period will result in the most unfavourable minimum value of $H_{D=0}$. The critical wave height derived from the weight of the block is a constant as the Hudson formula does not consider the wave period. In fact, when a certain wave period occurs, it is possible that a smaller wave height can destabilize a block with a calculated stable weight. In response to this scenario, a number of Dutch academics have proposed new and improved methods.

(2) The former Soviet union code methods. The СНиП 57-75 methods are provided below.

When designing a sloping structure consisting of blocks of stone, concrete or reinforced concrete, the individual constructive weight W, corresponding to its ultimate equilibrium state under wind waves, is calculated according to Eq. (4.5.21).

The block is located on a sloping section from the top of the structure to a water depth of $Z=0.7H$, then:

$$W=\dfrac{0.1\mu K_\gamma^3 \gamma_b H^2 \overline{L}}{\sqrt{1+m^3}} \quad (4.5.21)$$

Where, μ is a shape factor, determined by Table 4.5.17; K_γ is the heaviness factor, same as Eq. (4.5.16); γ_b is the heaviness of block material, kN/m^3; H is the wave height, m, adopted as wave height with cumulative frequency of 2%; $\overline{L}$ is the mean wave length, m; m is the slope.

Table 4.5.17　　　　　　　　　　Values for shape factor μ

Name of component	μ	
	Throwing	Layout
Blocks	0.0250	—
Universal concrete block	0.0210	—
Four-legged cone	0.0080	0.0058
Ferrous column	0.0057	0.0049
Triple column	0.0057	0.0034
Six-legged block	0.0043	0.0034

Eq. (4.5.21) reflects the effect of wave length. However, when the wave steepness is large, the block weight calculated by Eq. (4.5.21) is too large. Therefore, the Eq. (4.5.21) has been amended in the former Soviet Code (СНиП Ⅱ 2.06.04 – 82 *), published in 1986. The new equation is provided below.

$$W = \frac{0.316\mu K_\gamma^3 \gamma_b H^3}{\sqrt{1+m^3}} \sqrt{\frac{\overline{L}}{H}} \qquad (4.5.22)$$

When $Z > 0.7H$, the weight of a single block W_z shall be calculated according to the following equation.

$$W_z = W \exp\left(-\frac{7.5z^3}{H\overline{L}}\right) \qquad (4.5.23)$$

Where, W_z is the weight of the block calculated according to Eq. (4.5.22). The rest of the symbols are as before. Note that the units of W and γ_b in Eq. (4.5.21) and Eq. (4.5.22) are t and kN/m^3, respectively. Compared to the original equation, both equations are multiplied by a conversion factor of 0.1, because $1 \text{ kN} = \frac{1}{9.8} \text{ t} \approx 0.1 \text{ t}$. The unit of heaviness in the original standard is t/m^3.

When designing an armour consisting of unsorted blocks, a factor ε consisting of the grain size within the shaded area of Fig. 4.5.5 must be used.

$$\varepsilon = \sqrt[3]{\frac{W_i}{W} \frac{D_i}{D}} \qquad (4.5.24)$$

Where, W is the weight of blocks determined by Eq. (4.5.22) and Eq. (4.5.23), t; W_i is the weight of blocks larger or smaller than W, t; D_i and D are the diameters of the spherical blocks, responding to W_i and W, respectively.

The conditions for the application of Fig. 4.5.5 are $3 \leqslant m \leqslant 5$ and $H \leqslant 3$ m.

(3) Calculation of the thickness of the surface layer of riprap stone armour. For the riprap stone armour, it is generally considered that it should be 2 to 3 layers. In this way, local deformations of the slope can be automatically adjusted under wind waves, without causing extensive damage.

According to the design code for breakwater and seadike engineering, the thickness of the armour layer h(m) is calculated according to the following formula.

$$h = nc \left(\frac{W}{0.1\gamma_b}\right)^{\frac{1}{3}} \qquad (4.5.25)$$

Where, n is the number of the armour layers; c is a coefficient, as can be determined by Table 4.5.18.

The rest of the symbols are the same as before.

Number of artificial blocks:

$$N = Anc(1-P)\left(\frac{0.1\gamma_b}{W}\right)^{\frac{2}{3}} \qquad (4.5.26)$$

4.5 Design of seadike

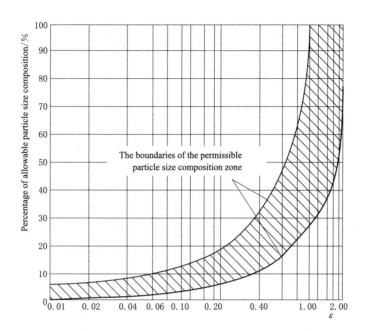

Fig. 4.5.5 Curves for determining the permitted particle size composition of unsorted ripraps for armours

Where, A is the average area of surface armour layer perpendicular to thickness h, m²; P is the porosity of the armour surface layer, %, as provided in Table 4.5.18.

Table 4.5.18   Values for c and porosity P

Armour blocks	Structural type	c	Porosity $P/\%$	Note
Blocks	Throwing 2 layers	1.0	40	
	Layout 1 layer (standing)	1.3 – 1.4	—	
Four-legged cone	Layout 2 layers	1.0	50	
I-block	Layout 2 layers	1.2	60	Random placement at fixed points
		1.1	60	Regular layout

3. Calculation of slab thickness of concrete armour

The design of concrete or reinforced concrete armour slabs should first be based on the buoyancy forces generated by the wave actions to determine the slab thickness. The reinforcement for the armour slabs placed on the flexible base is then calculated based on the wave breaking pressure. Only the slab thickness design is described below.

The whole concrete slab may float under the action of the floating support force. When the slope is 1 : 5 to 1 : 2, for concrete slabs with open joints, the Former Soviet Code СНиП Ⅱ 2.06.04 – 82 * the following formula for calculating the thickness required for stability (m).

$$t = 0.07CH^3 \sqrt{\frac{L}{S} \frac{\gamma}{\gamma_b - \gamma}} \frac{\sqrt{1+m^2}}{m} \qquad (4.5.27)$$

Where, C is the coefficient adopted as 1 for the integral armour slab and as 1.1 for assembled armour slabs; S is the side length of the armour slab in the direction perpendicular to the water edge line (i.e., along the slope), m; γ is the heaviness of water, kN/m^3; γ_b is the heaviness of armour plate, kN/m^3; $m = \cot\alpha$ is the slope with α of the angle between the slope and horizontal plain; L is the wave length, m; H is the wave height, m.

The maximum floating support force on the assembled slab near the water edge line is recommended to consider the effect of water flow with air. The formula for the thickness of the assembled slab $t(m)$, is as follows.

$$t = 0.66 \frac{\overline{H}^2 \sqrt[4]{(S/\overline{H})}}{S\cos\alpha} \frac{\gamma}{\gamma_b - 0.3\omega\gamma} \qquad (4.5.28)$$

Where, $\overline{H}$ is the averaged wave height, m; ω is the load saturation factor, a function of $S/\overline{H}$, as can be found in Table 4.5.19.

Table 4.5.19   Load saturation factor ω

$S/\overline{H}$	≤1.0	1.2 – 1.5	2.2 – 2.8	2.5 – 4.3	≥5.0 – 6.0
ω	1.0	0.8	0.7	0.6	0.5

Note: The rest of the symbols are the same as before.

4.5.3 Calculation of stability of protective wall

For vertical walls or straight-walled seawalls, if the protection wall is a gravity structure that relies on its own weight for stability, stability calculations should be carried out for the following aspects.

(1) In addition to the self-weight of the wall and the earth pressure behind the wall, the net pressure due to changes in tidal level and wave conditions in front of the wall should also be taken into account when calculating the wall's stability against overturning. Outside of the wall, high tidal level, low tidal level or tidal level at the top of the ballast should be used. Highest tidal level or the same as the tidal level outside of the wall should be used for the inside wall. Net wave pressure should be taken as the pressure at wave trough in front of the wall.

(2) Stability of the rear toe tilt means that the wall may fall landward during construction. In this case, the high tidal level during the construction period is taken outside of the wall and the corresponding low tidal level and backfill height inside the wall.

(3) Slip resistance of the wall as a whole along the base of the wall or along each horizontal joint of the wall.

(4) Slip resistance of the protective wall along the contact surface between the bedding and the foundation. The tidal level outside the wall is generally taken as the low tidal

level or the mudflat level, while the high tidal level is taken inside the wall when calculating the stability against tilting.

(5) Calculation of stability of foundation. Protective walls on steep-walled seadikes are sometimes not made into gravity structures, but rather dry or slurry stone armours, which should be calculated as the armour stability.

1) The wall overturning stability is calculated by the following formula.

$$K_0 = \frac{M_R}{M_0} \qquad (4.5.29)$$

Where, K_0 is the overturning stability safety factor, which should be determined according to the combination of construction level and load, generally $K_0 \geqslant 1.4 - 1.6$; M_R is the stabilizing moment for the front toe of the calculation domain, kN·m; M_0 is the overturning moment for the front toe of the calculation domain, kN·m.

2) The slip stability of each horizontal joint along the wall ground or wall body is calculated according to the following formula.

$$K_S = \frac{Gf}{P} \qquad (4.5.30)$$

Where, K_S is the slip stability safety factor which should also be determined according to the combination of construction level and load, generally $K_S \geqslant 1.1 - 1.3$; G is the horizontal combined force acting on the calculation domian, kN or kN/m; P is the horizontal combined force on the calculation surface, kN or kN/m; f is the friction coefficient along the calculation domian, as can be determined from Table 4.5.20.

Table 4.5.20   Values for friction coefficient

Materials	Friction coefficient
Concrete and concrete	0.55
Slurry block and slurry block	0.65
Block and block	0.70
Concrete and block	0.60
Slurry block and block (surface leveling with gravel)	0.65
Riprap and foundation with coarse and fine sands	0.50 - 0.60
Riprap and clay foundation	0.40
Riprap and subsand soil foundation	0.35 - 0.50
Riprap and clay, subclay foundation	0.30 - 0.45

3) The slip resistance of the protective wall along the bottom surface of the bedding layer is calculated according to the following formula.

$$K_S = \frac{(G+g)f + P_E}{P} \qquad (4.5.31)$$

Where, G is the vertical combined force acting on the bottom of the wall, kN or kN/m;

P is the horizontal combined force acting above the bottom of the wall, kN or kN/m; g is the material weight of bedding and ballast in the range of $CD-EE'$ (Fig. 4.5.6); P_E is the passive soil pressure on EE' surface, kN or kN/m. For a hidden foundation or the case shown in Fig. 4.5.6, 30% of the calculated value can be obtained.

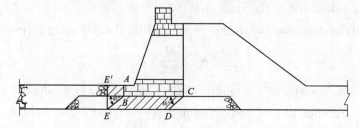

Fig. 4.5.6 Steep protective wall

4) For cohesive foundations, it is appropriate to use the following formula to calculate the slip stability along the surface of the foundation soil.

$$K_S = \frac{(G+g)\tan\varphi_0 + C_0 A + P_E}{P} \qquad (4.5.32)$$

Where, φ_0 is the friction angle between the substrate and the foundation, $\varphi_0 = \varphi$ when no measured data is available, φ is the repose angle of the foundation soil, (°); C_0 is the adhesive force along the sliding surface, $C_0 = \left(\frac{1}{4} - \frac{1}{6}\right)C$, where C is the cohesive force of foundation soil, kPa; A is the area of the bottom ($BC = ED$ in Fig. 4.5.6). The rest of the symbols are the same as before.

The above and C values can be used for the consolidation fast shear index of the indoor direct shear test.

5) Case. A steep wall seadike with gravity dry masonry protective walls has the cross-section dimensions shown in Fig. 4.5.7. The design high water is 3.7 m, mean tidal range is 4.4 m, design wave height is 1.7 m. Heaviness of block masonry $\gamma_1 = 22$ kN/m³, the friction factor with bedding layer $f = 0.6$. The heaviness of filling sea soil behind the wall $\gamma_1 = 17$ kN/m³, repose angle $\varphi_1 = 4°$, cohesive force $C_1 = 5$ kPa, heaviness of riprap $\gamma_2 = 17$ kN/m³, floating heaviness $\gamma' = 10$ kN/m³, repose angle $\varphi_2 = 38°$. Repose angle of the foundation soil (sea clay) $\varphi_2 = 13°$, cohesive force $C_{cu} = 2$ kPa. Please try to check the overturning and slip stability of the seaward side of the protective wall.

Solve: Checking the seaward side of the protective wall against overturning and slip stability, in principle, should consider the design of high and low tide wave force respectively (i.e., net wave pressure on the seaward side under the wave trough).

$$\therefore \quad \frac{d_1}{H_H} = \frac{2.2}{1.7} = 1.29$$

$$\therefore \quad \frac{d_1}{d} = \frac{2.2}{3.7} = 0.59$$

4.5 Design of seadike

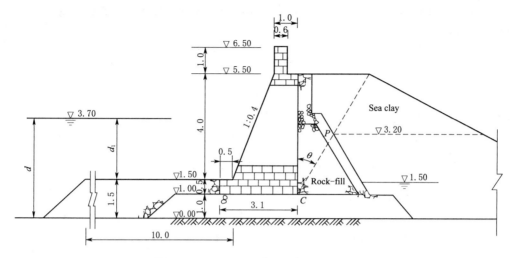

Fig. 4.5.7 Steep wall seadike (unit: m)

According to the *Code of Hydrology for Harbour and Waterway* (JTS 145—2015), the wave in front of the protective wall is in the near-break form in the middle foundation, and the wave pressure is very small and can be neglected.

Because of the average tide range is 4.4 m, the design low tidal level is lower than the flat elevation ±0, implying there is no wave force acting on the protective wall at the design low tidal level. The tidal level in front of the wall in this calculation is taken as 1.5, which is the same as the elevation of the ballast top. The tidal level behind the wall is assumed to be the same as outside the wall, without accounting for the remaining water pressure.

Checking the resistance to inclination and slip stability along the bottom plane of the wall.

a. Calculating the distribution of active soil pressure. Since there is a stone throwing body behind the wall, according to Section 2.2.11 of the Port Engineering Specification for Gravity Jetties, when the rupture surface passes through two types of filling, the soil pressure is calculated above and below the exit slope point P according to the indicators of the two types of filling respectively. The location of the P point is determined by approximation of $\bar{\theta}$ and $\bar{\theta}$ is the average rupture angle, which is obtained by weighting the rupture angles of the two types of fillings according to the corresponding layer thicknesses. In this case, the rupture angle of sea clay $\theta_1 = 45° - \frac{\varphi_1}{2} = 43°$, and $\theta_2 = 45° - \frac{\varphi_2}{2} = 26°$ for riprap behind the wall (stacking stones). $\bar{\theta}$ is an approximation of averaged values of θ_1 and θ_2:

$$\bar{\theta} = \frac{1}{2}(\theta_1 + \theta_2) = 34.5°.$$

Accordingly, the elevation of the exit point P can be found as 3.2.

For cohesive soil, when the ground is horizontal, the soil pressure P_a (kPa) on the back of the vertical wall or the calculated vertical surface is calculated according to the following formula.

$$P_a = \gamma h K_a - 2C\sqrt{K_a} \qquad (4.5.33)$$

$$K_a = \tan^2\left(45° - \frac{\varphi}{2}\right)$$

Where, γ is the heaviness of filling materials behind the wall, kN/m³; h is the thickness of the filling, m; K_a is the active soil pressure coefficient, for the sea clay, $K_a = \tan^2\left(45° - \frac{4°}{2}\right) = 0.87$; for the stacked stones, $K_a = \tan^2\left(45° - \frac{38°}{2}\right) = 0.24$; C is the cohesion of the soil, kPa, for the stacked stones, $C = 0$ kPa.

Since sea clay has cohesive force, part of the soil pressure is the tensile force caused by cohesion. It is independent of the depth from the ground, and is distributed in a rectangular shape along the wall height, with the maximum depth of the stretched area z_0 at soil pressure $P_a = 0$.

$$z_0 = \frac{2C\sqrt{K_{a1}}}{\gamma_1 K_{a1}} = \frac{2\times 5}{17\times\sqrt{0.87}} = 0.63 \text{ (m)}$$

The soil pressure calculation formula and calculation table for each different elevation behind the wall are shown in Fig. 4.5.8 and Table 4.5.21 respectively.

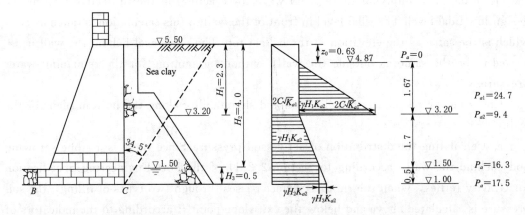

Fig. 4.5.8 Calculation of force distribution of steep wall seadike (unit: m)

Table 4.5.21   Background soil pressure

Elevation/m	Formula	Index	Pressure/kPa	Note
4.78	$P_a = \gamma z_0 K_{a1} - 2C\sqrt{K_{a1}}$	$\gamma = 17$ kN/m³ $z_0 = 0.63$ m $C = 5$ kPa $K_{a1} = 0.87$	0	The elevation is determined by z_0
3.20	For sea clay $P_a = \gamma H_1 K_{a1} - 2C\sqrt{K_{a1}}$	$H_1 = 2.3$ m γ, C, K_{a1}, the same as above	$P_{a1} = 17\times 2.3\times 0.87 - 2\times 5\times \sqrt{0.87} = 24.7$	

4.5 Design of seadike

Continued

Elevation/m	Formula	Index	Pressure/kPa	Note
3.20	For stacking stones $P_a = \gamma H_1 K_{a1}$	$\gamma = 17$ kN/m³ $H_1 = 2.3$ m $K_{a1} = 0.24$	$P_{a2} = 17 \times 2.3 \times 0.24 = 9.4$	
1.50	$P_a = \gamma H_2 K_{a2}$	$H_2 = 4.0$ m $\gamma = 17$ kN/m³ $K_{a2} = 0.24$	$P_a = 17 \times 4.0 \times 0.24 = 16.3$	
1.00	$P_a = \gamma H_2 K_{a1} + \gamma H_3 K_{a2}$	$\gamma = 10$ kN/m³ $H_3 = 0.5$ m C, K_{a2}, the same as above	$P_{a2} = 17 \times 4.0 \times 0.24 + 10 \times 0.5 \times 0.24$ $= 16.3 + 1.2 = 17.5$	

b. Calculation of total soil pressure and moment. The calculation process and results are shown in Table 4.5.22.

Table 4.5.22　　　　Calculation table of total pressure and moment

Pressure distribution figure	No.	Soil pressure /kN	Force arm on point B of the front toe /m	Moment to point B of the front toe /(kN·m)
	①	$\frac{1}{2} \times 24.7 \times 1.67 \approx 20.6$	$\frac{1}{3} \times 1.67 + 2.2 \approx 2.76$	$20.6 \times 2.76 \approx 56.9$
	②	$9.4 \times 1.7 \approx 16.0$	$\frac{1}{2} \times 1.7 + 0.5 = 1.35$	$16 \times 1.35 = 21.6$
	③	$\frac{1}{2}(16.3 - 9.4) \times 1.7 \approx 5.9$	$\frac{1}{3} \times 1.7 + 0.5 \approx 1.07$	$5.9 \times 1.07 \approx 6.3$
	④	$16.3 \times 0.5 \approx 8.2$	$\frac{1}{2} \times 0.5 = 0.25$	$8.2 \times 0.25 \approx 2.1$
	⑤	$\frac{1}{2}(17.5 - 16.3) \times 0.5 = 0.3$	$\frac{1}{3} \times 0.5 \approx 0.17$	$0.3 \times 0.17 \approx 0.1$
	Total	50.9		87

c. Calculating the self-weight and stabilizing moment of the protective wall. The calculation process and results are shown in Table 4.5.23.

Table 4.5.23　　　Self-weight and moment calculation table for per meter of wall width

	No.	Weight G_i/kN	Force arm on point B of the front toe l_i/m	Moment to point B of the front toe $G_{ai} l_i$/(kN·m)
	①	$0.5 \times 1.0 \times 22 = 11.0$	$2.1 + \frac{0.5}{2} = 2.35$	$11 \times 2.35 = 25.9$
	②	$1.0 \times 4.0 \times 22 = 88.0$	$2.1 + \frac{1.0}{2} = 2.60$	$88 \times 2.6 = 228.8$
	③	$\frac{1}{2} \times 1.6 \times 4.0 \times 22 = 70.4$	$0.5 + \frac{2}{3} \times 1.6 = 1.57$	$70.4 \times 1.57 = 110.5$
	④	$3.1 \times 0.5 \times 12 = 18.6$	$\frac{1}{2} \times 3.1 = 1.55$	$18.6 \times 1.55 = 28.8$
	Total	188.0		394.0

d. Checking the stability against overturning and sliding Stability against overturning. From Eq. (4.5.29):

$$k_0 = \frac{M_R}{M} = \frac{\sum_{i=1}^{4} G_i l_i}{\sum_{i=1}^{5} P_{ai} h_i} = \frac{394}{87} = 4.53 > 1.6$$

Stability against sliding. From Eq. (4.5.30):

$$K_S = \frac{Gf}{\sum_{i=1}^{5} P_{ai}} = \frac{0.6 \times 188}{51} = 2.21 > 1.3$$

e. Checking the slip stability of foundation soil surface. From Eq. (4.5.32):

$$K_S = \frac{(G+g)\tan\varphi_0 + C_0 A + P_E}{P}$$

Where, G, P are calculated as above.

$$\varphi_0 = \varphi_{cu} = 13°$$

$$C_0 = \left(\frac{1}{4} - \frac{1}{6}\right) C_{cu} = \frac{1}{5} \times 2 = 0.4 \text{ (kPa)}$$

Where, A is the unit-width wall bottom area, $A = 3.1 \times 1.0 = 3.1$ (m^2); g is the bedding and ballast weights for unit-width dike lengths within the range of $ABCDEE'$.

From Fig. 4.5.9:

$$g = \left[3.1 \times 1.0 + \frac{1}{2} \times (0.5 + 1.5) \times 1.0\right] \gamma' = 4.1\gamma' = 41 \text{ (kN)}$$

P_E is the 30% of unit-width passive soil pressure in domain EE'.

$$P_E = \frac{1}{2}\gamma' H^2 K_p \times 30\% = \frac{1}{2} \times 1 \times 10 \times 1.5^2 \times \tan^2\left(45° - \frac{38°}{2}\right) \times 30\% = 0.803 \text{ (kN)}$$

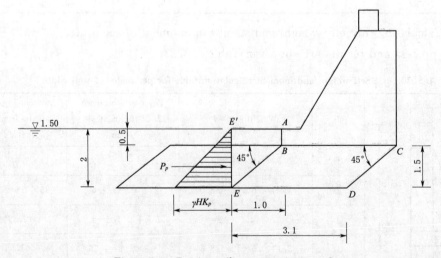

Fig. 4.5.9 Passive soil pressure (unit: m)

Substituting each of the above calculated values into Eq. (4.5.32), we get:
$$K_S = \frac{(188+41)\tan 13° + 0.4 \times 0.31 + 14.2}{51} = 1.32 > 1.3$$

The check results can meet the requirements of overturning and slip stability.

4.5.4 Stability calculation of wave mitigation breakwater

If the bottom elevation of the breakwater at the top of a sloping seadike is within the range of wave runup, the breakwater will be subjected to wave forces, and the calculation of breakwater stability is required. The wave horizontal force and uplift force are the main external loads on the breakwater. If the breast wall is designed to be buried into the top of the embankment, there is also passive soil pressure. However, if the burial depth is not large, the passive soil pressure can be neglected.

Wave pressure calculations for breakwater on the sloping seadike can refer to the Port and Waterway Hydrology Specification. The wave force on the breakwater of the upright or vertical seadike should firstly be determined according to the water depth in front of the dike and the water depth at the top of the foundation in accordance with the ratio of water depth to wave height to determine the wave form in front of the straight wall. Then, according to the wave state is a standing wave, far-breaking wave or near-breaking wave to choose the corresponding wave pressure calculation formula, and intercept the action of the wave pressure on the part of the breakwater as the wave calculation load. The specific calculation method is shown in Chapter 7 of the harbor engineering specification.

The stability calculation of the breakwater also includes two parts: overturning stability and sliding stability. Calculated tidal level takes the design high tidal level in front of the dike. Specific reference can be made to the principles and methods in Section 4.5.3.

4.5.5 Seadike slip resistance stability calculation

Seadikes usually built on soft clay foundations are often damaged by landslides due to the instability of the foundation. Soft clay foundation instability is prefaced by the following signs.

(1) Dramatic increase in subsidence.

(2) Dramatic increase in lateral deformation of the foundation causes the toe of the slope to bulge.

(3) Longitudinal cracks in the direction of the main axis appear in the dike, and their length and width gradually expand, followed by the development of transverse cracks, and finally develop into a sliding surface in the form of a circular curve.

(4) Rapid increase in pore water pressure in the foundation or with water infiltration at the toe of the slope.

These signs appeared as early as 4 – 5 days before the sliding damage of the seawall soil, which indicates that the destabilization of the dike body and foundation is a process of developing from local damage to overall damage.

The seadike foundation stability analysis mainly addresses two aspects.

(1) According to the measured strength index of foundation and the safety factor required by the project, design a reasonable seadike section to meet the requirements of service.

(2) For the existing seadike section size and foundation strength, the stability of the foundation is checked. The stability analysis of seadike foundation is done by circular sliding analysis.

There are various circular sliding methods, but no matter which method is used, the external load acting on the seadike must be analyzed first, considering the combination of different tidal levels inside and outside the dike, and different calculation situations such as the completed section and construction section should be analyzed.

When calculating the self-weight, the submerged part is calculated according to the floating weight, and the above-water part for the stacked stone is calculated according to the dry weight. The soil body can be calculated using saturation weight or wet weight. If there is a tidal level difference between inside and outside the seawall to produce infiltration, the effect of water infiltration pressure should generally be handled as follows. When calculating the sliding moment, the saturation weight is used below the infiltration line and above the design low tidal level, but the floating weight is used in calculating the anti-sliding moment. As for the location of the infiltration line, it can be simplified by connecting the intersection of the internal and external tidal levels and the slope of the impermeable soil with a straight line. For the unclosed stacked stone interceptor dike, the intersection of the internal and external tidal level and the slope of the interceptor dike can be connected by a straight line.

There are four main types of stability calculations determined by different design stages and different tidal level combinations.

(1) Preliminary design phase, stability calculations of the completed cross-sections.

(2) Preliminary design phase, stability calculation of the plugging cutoff section.

(3) Stability calculations for the non-blocked section of the seadike when the plugging is first completed during the construction design phase.

(4) Stability calculation of seadike construction sections during blockage in the construction design phase.

The above calculations are listed in Table 4.5.24.

Different calculation methods of circular sliding use different indicators of foundation strength.

General seadike stability analysis using total stress method. For the construction period section check, no drainage shear or fast shear strength index should be used. Section stability during the completion period can be checked by using solidification without drainage or solidification fast shear strength index. However, the calculation should be based

4.5 Design of seadike

on the actual situation of the slip resistance moment and strength indicators of the lifting body to make appropriate discounts.

Table 4.5.24   Calculation situations

Situations	Section characteristics	Calculation content	Tidal level combinations		Load combinations	
			Offshore	Inner harbor	Basic	Spectial
1	Completed domain	Slide outward	(1) Design low tidal level (2) Tidal level down to top of ballast	Highest flooding level	Self-weight infiltration force	Earthquake force
		Slide inward	Mean high tidal level	Lowest control tidal level		
2	Construction period domain	Slide outward	(1) Design low tidal level at integration period (2) Tidal level down to top of ballast	Highest tidal level at integration period	Self-weight infiltration force	
		Slide inward	Design high tidal level of interceptor dike	Lowest tidal level at integration period		
3	Construction period domain	Slide outward	(1) Design low tidal level at construction period (2) Tidal level down to top of ballast	Design highest tidal level at construction period	Self-weight infiltration force	
		Slide inward	Mean high tidal level at construction period	Design lowest tidal level at construction period		
4	Blocking section at the end of construction period	Slide outward	(1) Design low tidal level at construction period (2) Tidal level down to top of ballast	Same as the outer tidal level or depending on the situation	Self-weight infiltration force	
		Slide inward	Same as the inner tidal level	Design lowest tidal level at construction period		

Consolidation effective stress method should be used for large and medium-sized projects. Since the consolidation effect caused by the dike load has been taken into account in the calculation formula, the foundation strength index uses the nondrainage shear or consolidation nondrainage shear.

The $\varphi=0$ method is used for small-scale projects, and the strength index is determined by cross-plate shear on the field, and its strength increases with the depth of the foundation.

Large projects can also use the effective stress method, or the simplified Bishop method, when available. The method should use the effective strength index and accurately determine the pore water pressure distribution. It is generally measured by triaxial consolidation undrained shear test which measures pore water pressure, and straight shear instrument for slow shear test can also be used. It is better to be checked the field measurements of actual pore water pressure. For the case of soft soil interlayer in the foundation, it is appropriate to use the simplified method of composite sliding surface to calculate.

The coefficient of safety for the seadike slip resistance and stability should take into account the calculation method, strength index and structure grade, load combination, etc., generally selected according to Table 4.5.25.

Table 4.5.25   Coefficient of safety for seadike slip resistance and stability

Calculation method	Consolidated effective stress method	$\varphi=0$ method	Effective stress method (Simplified Bishop method)
Safety effect	1.05 - 1.25	1.1 - 1.3	1.3 - 1.5

The following are a list of the main methods of calculating circular sliding respectively.

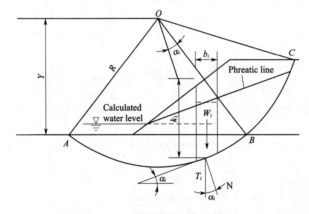

Fig. 4.5.10   Total stress method

(1) Total stress method. As it can be seen in Fig. 4.5.10, the anti-slip moment M_R (kN·m/m) and sliding moment M_S (kN·m/m) for circular sliding are:

$$M_R = (\sum C_i l_i + \sum W_i \cos\alpha_i \tan\varphi_i) R$$
(4.5.34)

$$M_S = (\sum W_i \sin\alpha_i) R$$
(4.5.35)

Where, l_i is the arc length of the i-th soil bar, m; W_i is the self-weight of the i-th soil strip, kN/m, $W_i = \gamma_i b_i h_i$, γ_i is the heaviness of the soil trip. Between the immersion line and the tidal level line, saturation weight is used to calculate the sliding moment and floating weight is used to calculate the anti-slip moment. Floating weights are used below the tidal level; b_i and h_i are width and mean height of the soil strip, m; α_i is the angle between the tangent to the horizontal line at the midpoint of the i-th soil strip arc, (°); R is the radius of the slip arc, m; C_i, φ_i are indicators of shear strength on the sliding surface of the i-th soil strip, kPa, (°).

Index of undrained shear strength of foundation soils should be selected on the AB arc section and dike soil consolidation undrained shear strength index should be selected on BC arc section.

4.5 Design of seadike

Shear strength of soils is

$$\tau_i = C_i + \sigma_i \tan\varphi_i$$

Where, σ_i is the normal pressure on the sliding surface.

The total stress method assumes that the sliding soil is a rigid body without deformation and does not take into account the forces acting on both sides of the soil strip. It should be noted that when the center of the base of the soil strip is at the right of the vertical line of the center of the sliding arc, the sliding shear force is in the same direction as the sliding and should take a positive sign. The shear force is opposite to the sliding direction on the left side of the vertical line and should be taken as a negative sign to act as a shear resistance.

The sliding stability factor of safety K is:

$$K = \frac{M_R}{M_S} \tag{4.5.36}$$

The above derived K is the sliding safety factor for an arbitrary assumed sliding surface. It is usually necessary to assume a series of sliding surfaces and their centers for several trial calculations to find the minimum safety factor corresponding to the most dangerous sliding surface.

(2) Consolidated effective stress method. The sliding stability factor of safety K for the consolidation effective stress method is calculated according to the following formula:

$$K = \frac{\sum\limits_{A}^{B}(C_u + W_{\text{I}i}\cos\alpha_i \tan\varphi_u + W_{\text{II}i}U\cos\alpha_i \tan\varphi_{cu})}{\sum\limits_{A}^{B}(W_{\text{I}i} + W_{\text{II}i})\sin\alpha_i + \sum\limits_{B}^{C}W_{\text{II}i}\sin\alpha_i} + \frac{k_1 \sum\limits_{B}^{C}(C_{\text{II}}l_i + k_2 W_{\text{II}i}\cos\alpha\tan\varphi_{\text{II}})}{\sum\limits_{A}^{B}(W_{\text{I}i} + W_{\text{II}i})\sin\alpha_i + \sum\limits_{B}^{C}W_{\text{II}i}\sin\alpha_i} \tag{4.5.37}$$

Where, $W_{\text{I}i}$ and $W_{\text{II}i}$ are the weights of the i-th soil strip in the foundation part and the dike part, kN/m; U is the consolidation of the foundation soil; C_u and φ_u are the indexes of undrained shear strength of foundation soils, kPa, (°); φ_{cu} is the angle of repose of consolidated undrained shear foundation soils, (°); C_{II} and φ_{II} are shear strength indexes of the dike, derived from undrained shear, kPa, (°); k_1 is the reduction factor for the slip moment of the dike; k_2 is the reduction factor for dike strength index; the rest of the symbols have the same meaning as before.

The calculation diagram is shown in Fig. 4.5.11.

When the construction period is long and the foundation is partially consolidated by the load of the dike, it is advisable to use the consolidated effective stress method. If the consolidation of the foundation is not considered, then $W_{\text{II}i}U\cos\alpha_i\tan\varphi_{cu}=0$ and $W_{\text{I}i}\cos\alpha_i\tan\varphi_u$ in Eq. (4.5.37) should be changed into $W_{\text{I}i}+W_{\text{II}i}\cos\alpha_i\tan\varphi_u$.

In this way Eq. (4.5.37) becomes the total stress method taking into account the sliding moment of the dike and the strength index reduction factor. Since the dike body soil

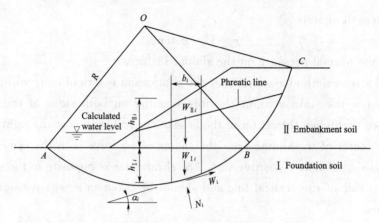

Fig. 4.5.11 Consolidated effective stress method

has lost or partially lost its slip resistance during foundation sliding, the slip moment caused by the dike body soil can often be disregarded or only partially considered in the seadike stability analysis. Thus a reduction factor k_i is used in Eq. (4.5.37), generally adopted as 0.6 – 0.8.

(3) $\varphi = 0$ method. $\varphi = 0$ method is often used in small-scale reclamation projects. It uses a cross-plate strength index that reflects the actual strength of the soft foundation, but cannot reflect the change in strength due to load changes. Cross-plate strength increases with depth:

$$\tau = C_0 + \lambda z$$

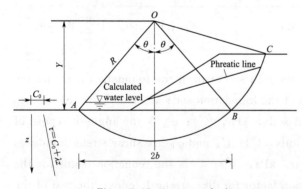

Fig. 4.5.12 $\varphi = 0$ method

Where, C_0 is the intercept of the cross-plate strength-to-depth relationship curve; λ is the rate of the cross-plate strength-depth relationship curve; z is the depth from the surface of the foundation, m.

The calculation diagram is shown in Fig. 4.5.12.

Slide stability factor of safety can be used according to Eq. (4.5.38).

$$K = \frac{2R[(C_0 - \lambda y)\theta + \lambda b] + k_1 \sum_{B}^{C}(C_i l_i + k_2 W_i \cos\alpha_i \tan\varphi_i)}{\sum_{C}^{A}(W_i \sin\alpha_i)} \quad (4.5.38)$$

Where, θ is the 1/2 of the angle of the center of the circle corresponding to the arc of AB, rad; y and b can be found in Fig. 4.5.12; the rest of the symbols are as before.

(4) Effective stress method (simplified Bishop's method). The Bishop's method is also a type of circular sliding method, which is characterized by the consideration of the

4.5 Design of seadike

forces acting on the sides of the soil strips. In addition, it uses an effective strength index, which requires the determination of pore water pressure.

The factor of safety for overturning stability by the simplified Bishop method is calculated by Eq. (4.5.39).

$$K = \frac{\sum \frac{1}{m_{ai}}[C'_i b_i + (W_i - u_i b_i)\tan\varphi'_i]}{\sum W_i \sin\alpha_i} \quad (4.5.39)$$

Where, C'_i and φ'_i are effective strength indexes on the sliding surface of the i-th soil strip, kPa, (°), measured by triple-assisted consolidation undrained shear test with pore water pressure measurement, or by slow shear with a straight shear gauge; u_i is the pore water pressure on the sliding surface of the i-th soil strip, kPa; b_i is the width of the i-th soil strip, m; $m_{ai} = \cos\alpha_i + \frac{\sin\alpha_i \tan\varphi'_i}{K}$; the rest of the symbols are the same as before.

The sliding stability factor K is included in m_{ai} and therefore requires a trial calculation. For trial calculation purposes, Fig. 4.5.13 can be used.

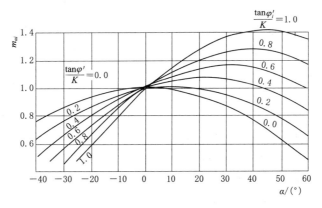

Fig. 4.5.13 Curves for m_{ai}

First assume a value of K, generally $K = 1$. According to K, α_i and $\tan\varphi'_i$, with m_{ai} found from Fig. 4.5.13, then K is calculated by Eq. (4.5.39). If the found K is different from the assumed K value, then check the m_{ai} graph again to find the new calculated K value based on the found K. This is repeated several times until the difference between the two K values is less than the permissible error.

The effective stress method is more reasonable, but the difficulty is that it requires a high level of instrumentation to test the strength index and is not easy to promote. When using this method, it is important to note that when the α_i value is negative, whether m_{ai} converges to zero. If this were the case, the effective normal reaction force N'_i acting on the soil strip would tend to infinity, which is clearly unreasonable. Therefore, when the value of any soil strip $m_{ai} \leqslant 0.2$, it is advisable not to use this method to avoid large errors in the derived K values. In addition, when the α_i value of the soil strip at the top of the slope is large, this will also result in a negative value of the effective normal reaction force N_i' of the strip, which is also unreasonable, and this reaction force can then be taken to be zero. This means that the shear force on the sliding surface of the soil strip has only an effective cohesive effect and without an effective internal friction angle effect. The Eq. (4.5.39) can be simplified as:

$$K = \frac{\sum \dfrac{C_i' b_i}{\cos\alpha_i}}{\sum W_i \sin\alpha_i} \qquad (4.5.40)$$

(5) Sliding surface sliding stability factor of safety method. The coefficient of safety method for sliding surface sliding stability can be calculated in accordance with Eq. (4.5.41).

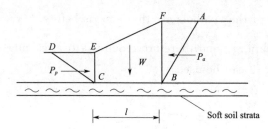

Fig. 4.5.14 Sliding surface stability force analysis diagram

$$K = \frac{W\tan\varphi + Cl + P_E}{P_a} \qquad (4.5.41)$$

Where, C and φ are the shear strength indexes for weak interlayers; P_a and P_E are active and passive earth pressures respectively; other symbols are shown in Fig. 4.5.14.

This method is only a simplified calculation method, can only be used for preliminary estimates, detailed calculations should refer to the relevant methods and regulations in the *Hydraulic Design Manual*.

4.5.6 Foundation subsidence calculations

Subsidence of seadike foundations is mainly the volumetric deformation of the soil caused by the self-weight and external loads of the seadike and the positive stresses within the foundation. Excessive foundation subsidence not only affects the normal use of the structure, but often leads to its destruction. For example, excessive subsidence of the seadike may cause cracks in the body of the dike, which may increase infiltration and affect the safety of the seadike.

Seadike foundation subsidence calculations are mainly performed for soft foundations. Silt, silty soils and generally clayey soils with low natural strength, high compressibility and high water content are collectively referred to as soft soils. Clayey soils with a natural moisture content greater than the liquid limit and a natural pore ratio greater than or equal to 1.5 are known as silt. A clayey soil with a natural porosity ratio of less than 1.5 but greater than or equal to 1.0 is called a silty soil.

The purpose of calculating the subsidence of foundations is to control the amount of subsidence within the permissible limits. It is also used to estimate the amount of additional earthwork or the amount of extra height to be reserved for fill due to subsidence.

Foundation subsidence mainly consists of initial subsidence and consolidation subsidence. Initial subsidence is the part of the subsidence that occurs immediately after the foundation load has been applied. For saturated foundations, this is due to the lateral expansion of the soil; for unsaturated foundations, in addition to the lateral expansion of the soil, it is also due to the compression or discharge of the pore gas between the soil particles.

If the base width of the seawall foundation is relatively large in relation to the thickness of the compression layer of the foundation, the initial subsidence can be neglected, but if this relative ratio is small, the initial subsidence cannot be ignored.

Consolidation subsidence is the main part of the total foundation subsidence and is caused by the discharge of water and air from the pores of the foundation under load, causing the volume of the soil to slowly compress and become smaller.

In fact, there is also so-called sub-consolidation subsidence in the later stages of soil consolidation, which is of such a long duration that it can generally be neglected.

1. Initial subsidence

The initial subsidence S_i can be calculated as follows:

$$S_i = \rho p \frac{B(1-\mu)^2}{E} \quad (4.5.42)$$

Where, p is the homogeneous pressure of seadike substrate, kN/m^2; B is the width of the short side of the foundation, m; μ is the the Poisson's ratio of the soil (for saturated clay, $\mu = 0.5$); E is the modulus of elasticity of foundation soil, kPa, generally obtained from triaxial undrained shear test or uniaxial compression test. It is best to perform repeated loading and unloading tests in the triaxial apparatus to find out the reloading modulus, which is used as the modulus of elasticity; ρ is the influence factor, referring to the provisions of the technical service for the design of reclamation projects in Fujian Province, the values can be taken according to Table 4.5.26.

Table 4.5.26   Values for ρ

Aspect ratio of the foundation	Flexible Foundation			Rigid foundation
	Values for ρ			Value
	Center point	Angle point	Average of full basis	Average of full basis
$L/B=2$	1.53	0.77	1.30	More flexible than slightly smaller foundation
$L/B=3$	1.78	0.89	1.52	
$L/B=5$	2.11	1.05	1.83	
$L/B=10$	2.58	1.29	2.25	
$L/B=100$	4.00	2.00	3.70	

If E value data is not available, it can be estimated according to Eq. (4.5.43).

$$S_i = \left(\frac{1}{4} - \frac{1}{3}\right) S_0 \quad (4.5.43)$$

Where, S_0 is the total subsidence when the side piles stop moving outward after the loading is completed.

2. Consolidation subsidence

Consolidation subsidence S_c is calculated according to Eq. (4.5.44) based on the sum-of-layers method.

$$S_c = \sum S_j = \sum \frac{e_{1j} - e_{2j}}{1 + e_{1j}} h_j \qquad (4.5.44)$$

Where, e_{1j} is the porosity when the pressure has been stabilized under the self-weight stress of the i-th layer of soil; e_{2j} is the porosity of the j-th layer of soil when compressed and stabilized by the combined effect of self-weight stress and additional stress; h_j is the thickness of the j-th layer of soil, m; S_j is the compression of the j-th layer of soil, m.

The calculation steps for the sum of layers method are:

(1) Determining the subsidence calculation domain. Selecting a number of subsidence calculation points on each cross-section. Generally, a number of control points on the midline of the dike axis and the vertical lines on both sides should be selected, and the location of the points should be determined by combining the properties of the foundation soil layer. The magnitude and distribution of the substrate stresses are then found according to the characteristics of the load (central or eccentric, etc.).

(2) Dividing the foundation into layers, the intersection of natural soil layers and the water table should both be divided into layers. The thickness of each layer should not be too large, generally 2 - 4 m, or equal to 0.4 times the width of the foundation. The total calculated thickness is determined by the depth at which the vertical additional stress is equal to 0.1 - 0.2 times the vertical self-weight stress.

(3) Calculating the vertical self-weight stress σ_S (starting from the ground) at each layer (0, 1, 2 in Fig. 4.5.15) on the vertical line of the control point. Giving its distribution curve, see Fig. 4.5.15.

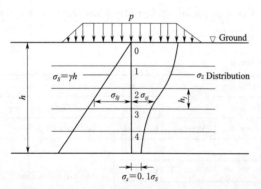

Fig. 4.5.15 Vertical additional stress diagram for each level of the sum-of-layers method

(4) Calculating the vertical additional stress at each layer on the vertical line of the control point. Starting from the foundation, when the foundation has a burial depth d, the distributed load strength needs to be subtracted from γ_d before calculating the additional stress, γ is the heaviness of the soil. Plotting its distribution curve. For the σ_z under the vertical load of the trapezoidal distribution of the strip foundation of the seawall section, it can be calculated according to the following formula.

$$\sigma_z = k_q p \qquad (4.5.45)$$

Where, p is the maximum trapezoidal load; k_q is the additional stress factor, check Fig. 4.5.16 to determine.

If there is a horizontal load on the foundation, the vertical stress caused by the horizontal load will be added to σ_z iteratively. Similar additional stress coefficients for other types of distributed loads can be found in the charts of the relevant specifications.

(5) Calculating the average self-weight stress σ_{Sj} and average additional stress σ_{Sj} of each layer by arithmetic average. Find out the porosity e_{1j} and e_{2j} corresponding to σ_{Sj} and $\sigma_{Sj} + \sigma_{zj}$ respectively from the ballast curve ($e - p$ curve or $e - \log p$ curve) of the foundation soil.

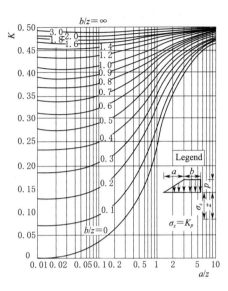

Fig. 4.5.16 Additional stress factor under trapezoidal distribution pressure

(6) Calculating the sum of subsidence of each layer S_C according to Eq. (4.5.45). The final subsidence is $S_\infty = m_S S_C$, m_S is the empirical factor. Because the sum-of-layers method is based on the formula for calculating the compression volume under the condition of no lateral deformation. m_s reflects additional subsidence caused by lateral plastic deformation, related to construction speed, etc., generally 1.2 – 1.6. Subsidence of soft ground is a long-term process. It often takes several years or even decades to complete all the subsidence after the seadike project is completed, and this feature needs to be fully noted when designing the seadike.

4.5.7 Soft ground reinforcement

China's coast is mostly silty coast, so the seadike foundation is often a soft foundation with the mud thickness of a few meters to a dozen meters, or even more than 20 m. Due to the low natural strength, high water content, high compressibility and poor permeability of silt and silty soils, various corresponding foundation reinforcement measures must be taken.

According to the principle of foundation, reinforcement soft foundation treatment methods can be divided into the following categories.

1. Displacement sand layer method

Replacement sand layer method also known as dredging method, that is, the use of dredger will be part of the foundation of all the said mud removed and replaced with sand and gravel layer. The method is generally suitable for the case where the thickness of the replacement filling is not more than 4 m. If the silt layer is soft and thick and cannot be removed completely, it can also be taken to remove the silt to a certain depth and then backfill with sand and gravel material. However, the displacement will still leave a thick layer of silt. To ensure the stability of the foundation, other foundation reinforcement methods are often needed. To improve the stability and strength of the foundation, one can set up ballast platforms on both sides of the building or lay drainage sand wells and plastic drainage boards in the soft soil layer for drainage pre-pressure consolidation.

2. Suppression layer method

When the thickness of soft soil layer is not suitable for dredging method, the suppression layer, also called suppression platform, can be filled on one or both sides of the seadike at the foot of the dike. Its function is to increase the anti-slip moment of the sea block section, improve the safety factor of circular sliding, and also improve the stress distribution of the dike foundation, which is beneficial to the stability of the foundation. The width and thickness of the suppression layer of the seadike should be determined by stability analysis calculations. First assume the size of the suppression layer, then change its size to meet the stability requirements according to the circular sliding method for trial calculation. As an estimate, the thickness (or height) of the suppression layer can be half of the dike height, and its width can be 2.5 - 3 times of the dike height, usually about 6 - 20 m. The ultimate height of the dike can be increased with the use of ballast layer measures, but not more than 1.6 - 2 times the ultimate height of the dike allowed by the foundation without ballast layer.

The suppression layer method requires a larger amount of soil and stone and increases the subsidence of the foundation, which is generally more suitable for the construction of soil dikes and dams that can withstand a large amount of subsidence. The construction of the suppression layer should be carried out simultaneously with the dike body to prevent disturbing the foundation soil.

3. Drainage sand bedding method

Drainage sand bedding layer is a layer of drainage sand laid on the foundation surface to form a horizontal drainage channel between the dike body and the soft foundation. Its role is firstly to accelerate the consolidation of soft ground, because the drainage consolidation time of the soil is proportional to the square of the drainage distance. The presence of the sand bedding layer shortens the drainage distance and accelerates the consolidation of the foundation soil, thus increasing its shear strength more quickly. It can increase the stability of the foundation, make the dike body load more uniformly distributed to the soft foundation, and help prevent the damage caused by the local stress concentration of the foundation.

This method of installing a horizontal drainage sand bedding layer on the ground is generally used in cases where the thickness of the soft soil layer does not exceed 5 m. The determination of the thickness of the sand bedding layer is based on the lifting body load diffusion to the sand bedding layer and soft soil layer intersection at a certain angle, where the stress should meet the requirements of the loading capacity of the foundation soil. If

4.5 Design of seadike

the thickness of soft soil layer is above 5 m, vertical row of permanent pre-pressure consolidation method should be used.

4. Vertical drainage pre-pressure consolidation method

When the soft soil layer is thicker, the foundation drainage consolidation time is longer, in order to shorten the consolidation time, the drainage distance must be shortened, for this reason, vertical drainage channels can be laid at certain intervals, and then the surface of the foundation is covered with drainage sand bedding for ballasting, which can achieve the purpose of accelerating consolidation.

Vertical drainage channels can be constructed with different materials and methods such as drainage sand wells, bagged sand wells or plastic sheet drainage.

Drainage sand wells are formed by using pile-driving equipment to drive steel pipes into the soil, then filling them with sand and pulling them out to form sand wells. The diameter of sand wells is generally 20 – 30 cm, underwater is 30 – 40 cm, the spacing of sand wells is usually around 2 – 4 m, and the length should not exceed 20 m. The thickness of the drainage sand bedding layer at the top of the sand well can be 0.3 – 0.5 m on land and 1 m underwater. Fig. 4.5.17 shows an engineering example of using a drainage sand well, which is the section of Seadike No. 9 in Daguanban, Lianjiang, Fujian.

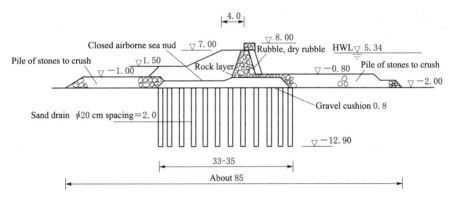

Fig. 4.5.17 Section of seadike No. 9 in Daguanban, Lianjiang, Fujian (unit: m)

Bagged sand well is a soft foundation reinforcement method developed on the basis of drainage sand well, which is made by putting sand into a permeable bag in advance and then into the borehole. This requires permeable bags of material with sufficient strength, good permeability, can play a role in filtering soil particles, usually using synthetic fiber polypropylene woven bags, which is more economical. The diameter of bagged sand well is generally 6 – 7 cm, 1 – 1.5 m spacing, 10 – 20 m length.

The cross-sectional area of plastic drainage board is generally 100 mm × 4 mm – 100 mm × 7 mm, which translates into a sand well diameter of 7 – 15 cm. It also has to be laid into the foundation by special machine using conduit with a spacing of 1 – 1.5 m. At present, the depth of the plastic plate is generally less than 20 m and up to 23 m. The main properties of plastic drainage board are the strength of the material, the longitudinal water circulation capacity and the soil barrier of the filter membrane. This requires good water circulation performance, low resistance to water circulation, and the need to meet the drainage process of retaining soil and not allowing oversized soil particles to invade the filter membrane and block the drainage channel. Since the quality of plastic drainage board is easy to guarantee, light weight and easy to construct, it overcomes the technical problems of sand wells which use a lot of sand and are prone to well shrinkage and broken wells which make the drainage channels fail, so it has been more widely used. The longitudinal water permeability of plastic drainage board is much better than that of bagged sand wells, so the effect of reinforcing soft foundation is also better, and it has been applied more often than bagged sand wells, and there is a trend of replacing bagged sand wells.

5. Geotextile laying method

When the silt layer is thicker, geotextiles can be laid on the dike body and foundation surface to play the role of anti-filtration, drainage, isolation and reinforcement to reduce uneven subsidence, reduce lateral deformation, and increase the stability of the foundation. On very soft foundation soil, the laid drainage sand bedding layer tends to fall unevenly into the soft foundation, so that the sand and silt mix, reducing the role of the sand bedding layer. At this time, one or two layers of geotextiles can be laid on the silt surface, which can play an isolating role to avoid mixing. For example, in the west breakwater of Chiwan Shenzhen, the thickness of the silt layer is 8 – 12 m, the local dredging replaced the sand bedding layer, and two layers of geotextiles were sandwiched in the sand bedding layer, and the 8 m high breakwater was successfully built. Another example is the ash storage of Beilun port power plant, the thickness of the foundation silt layer is 20 – 31 m, and the surface layer has 50 cm of liquid silt. After laying two layers of spun geotextiles on the surface, 40 cm of slag was laid on top of the geotextiles during construction, which can be put directly by car with the end-in method, speeding up the construction speed.

Geotextile applications do not need to consider strength reduction due to creep of the material because of its short-term stability during the construction period.

The friction between geotextile and the foundation soil can be calculated with the formula below.

$$F = \alpha \sigma \tan\varphi + \beta C \qquad (4.5.46)$$

Where, C and φ are the strength indexes of soil; α and β are friction and adhesion influence coefficients, respectively, which should be obtained by pulling and tensile tests of geotextiles in the figure. When there is no experimental data, $\alpha = 0.8$ and $\beta = 0$ can be adopted; σ is the vertical stress.

For the geotextile conditions under the arc sliding calculation, it should first find the minimum safety factor without geotextile sliding surface. Based on this, the additional slip resistance moment of the geotextile is taken into account, so that it meets the requirements of the design safety factor. Previously introduced Eq. (4.5.34) and Eq. (4.5.35) are respectively the anti-slip moment and sliding moment when not using geotextiles. When the geotextiles are considered, an additional term $(T_p \cos\zeta)R$ should be added to Eq. (4.5.34), T_p is the design tension for geotextiles and R is the angle of inclination between the intersection of the geotextile and the sliding arc and the center of the circle. Therefore, $(T_p \cos\zeta)R$ is the slip resistance moment added by the geotextile. The rest of the symbols are the same as in Eq. (4.5.34) and Eq. (4.5.35).

There are many other methods of soft foundation treatment, and the best reinforcement plan should be determined according to the building requirements, foundation conditions, construction conditions, material supply and construction period, etc., after a comprehensive technical and economic comparison. For example, Lianyungang West Embankment and Zhejiang Shengsi Tonggang Breakwater use the explosion dredging method, with an effective dredging depth of more than 16 m. The deep cement mixing method has been used in the east solid dike part of Tianjin Port, etc. Relevant information and specifications can be examined.

4.5.8 Seadike seepage prevention and blocking

The working condition of the seadike is that it is often under the pressure of infiltrating water formed by a certain level of water table difference. Seepage water pressure may not only take away the soil particles of the dike body forming "piping effect" but also cause the loss of the entire soil where the infiltration water escapes, resulting in the phenomenon of "soil flow". Moreover, it also increases the floating support force at the base of the structure, increasing the risk of sliding instability. Therefore, seepage control and blocking is an important component in the design of seadikes.

1. Seepage prevention design

The purpose of seepage prevention design is to ensure that no infiltration deformation occurs in the dike foundation. It consists of two parts. Firstly, it requires the foundation width of the dike to meet the requirement of greater than or equal to the specified seepage diameter length. Secondly, it is required that the weight of the soil at any point on the dike slope is greater than the infiltration water pressure acting at that point, i.e., the phenomenon of "soil flow" will not occur.

Let the bottom width of the seadike seepage prevention soil body is $L(m)$, Δh is the head difference (m) generated by the tidal level on both sides of the seadike, generally take the outer design high tidal level, while the inner side takes the design low tidal level, then:

$$L \geqslant c \Delta h \tag{4.5.47}$$

Where, c is the coefficient of seepage diameter.

The values of c can be selected according to the Table 4.5.27.

Chapter 4 Seadike Engineering

Table 4.5.27 Seepage diameter coefficient

Soil type	Value	
	Without back filtration layer	With back filtration layer
Silt	12	8
Clay	12	8
Fine sand	10	6
Medium, coarse sand	8	5
Clay loam	7	4

In order to ensure that "soil flow" does not occur, the weakest possible part of the impermeable soil of the dike body needs to be tested to meet the requirements of Eq. (4.5.48).

$$y = k \frac{\gamma \Delta h}{\gamma_b} \qquad (4.5.48)$$

Where, γ and γ_b are the water and soil heaviness respectively; Δh is the action water head, m, the difference between the design high tidal level on the outer sea side and the elevation of the check point is used at each point above the inner tidal level; The points below the inner tidal level are equal to the difference between the design high tidal level and the inner measured tidal level; y is the shortest distance from the test point to the soil surface, see point A in Fig. 4.5.18; k is the safety factor, generally greater than or equal to 1.2.

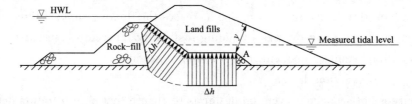

Fig. 4.5.18 Difference in water head imposed on impermeable soil

Eq. (4.5.48) indicates that the weight of the soil layer at any point $\gamma_b y$ is greater than the water pressure acting at that point $\gamma \Delta h$. The seepage prevention soil is generally filled with sea clay, but mountain soil (i.e. loess) is also used. The water content of mountain soil is small, so it is not easy to produce cracks after drying and has good impermeability after compaction. However, it is easy to be washed by the tide, so it is better to use sea clay below the high tide level, and then use mountain soil above. When using mountain soil, the main surface of the slope needs to be covered with sea clay in order to protect the slope surface. There should be no stones or slag mixed in the soil when filling with mountain soil, and require densely compacted.

To prevent the loss of seepage prevention soil under the action of waves and seepage flow, a backfilter layer should be installed between the seepage prevention soil and the rock pile or with the armour. The backfilter layer requires a certain strength of the material. Crushed stone, gravel, coarse sand, etc. can be used as backfilter layer material. If necessary, the design of the backfilter layer can be carried out in accordance with the aforementioned requirements for the retaining armour layer.

2. Blocking method

Leakage problems in the seadike will not only affect the normal use of the seadike, but will also endanger the safety of the dike. Leakage can often be caused by several conditions.

(1) In the soil-stone mixture with two internal and external stone prisms, the middle of the double prism is an impermeable core wall. If the improper positioning of riprap in water causes double prismatic collusion, it will cause leakage.

(2) Mountain soil dike is not adequately rolled or poor quality of soil material, forming a natural separation layer during construction, and the bottom layer of stone overflow is deposited.

(3) Insufficient seepage diameter length of seadikes with drainage bedding.

(4) The protection of the head wrapping at both ends of the mouth door of the wall mouth section is not removed cleanly, etc.

The following measures are commonly used for blocking leakage.

(1) Clay cover. The thickness of the cover should meet the requirement of not producing "soil flow", and not less than 1 m. Cover width according to the edge of the slope net wet treatment area, and then extend 3 - 6 times the head of water to determine, to prevent seepage flow escape at the transfer. Fujian East Lake pond dike leakage up to 0.7 m/s, had used this method to deal with.

(2) Clay interceptor wall. Dredge and backfill clay along the inner foot of the dike in the direction of the dike axis to become an interceptor wall. The general requirement is that the bottom width is not less than 1 m. The bottom elevation is dug to 0.3 - 0.5 cm in the impermeable layer, the length of the cut-off wall should be 10 - 20 m longer than the edge of the treatment area, and the wall height is about 2 - 3 m. This method has been used in Liudou Polder to treat leakage in the blocked section.

(3) Pressure sand filling. Use pressure sand filling to fill the pore space of rock pile, which can achieve the purpose of reducing leakage. Generally, if the sand filling volume reaches 35%-40% of the volume of the rock pile, the leakage volume will be reduced by about 30%. Control the sand filling pressure at 1 - 3 kg/cm^2, and gradually increase the pressure to prevent sand bubbling. The seawall of Donghutang has been tested with 8 holes of sand filling, with a well distance of 10 m, depth of 3.8 - 5.5 m, pressure of 0.4 - 0.6 kg/cm^2, cumulative filling time of 21 - 41 h, and sand filling volume of 21 - 98 m^3 per hole.

(4) Decompression wells. Decompression well is a kind of hollow surrounding well to deal with dike base overturning sand gushing water. The water storage around the well is

used to raise the water level of the inner seepage escape area, to reduce the seepage head difference, so that the water at the hole gush out without sand. It is suitable for the situation where the seepage head is not large and the seepage escape area is concentrated. At this time, the well pad of the perimeter well is not stacked high, and the stability of the well pad itself is not a problem, so using pressure reduction and regulating the water level inside and outside is an effective engineering solution.

Seawall leakage blocking is a kind of mending measure. Seepage prevention methods should be actively taken in the design and construction, which is the fundamental way to eliminate leakage from the dike.

4.6 Ecological seadike

4.6.1 Ecological seadike concept

At present, there is no unified and clear definition of the concept of ecological seadike in China, but relevant scholars have summarized some experiences in various studies and engineering practices.

Most domestic scholars agree that ecological seawalls should meet at least three major functions: physical, ecological and cultural. Firstly, physical resistance, disaster mitigation and prevention are the most important functions of seadikes. Secondly, we should try to preserve or artificially create vegetation and landscape elements close to nature, and maintain the original basic process of maritime life in the sea as far as possible. Thirdly, it should meet the needs of local leisure, entertainment and science education. Of course, for the specific manifestation of the specific ecological seadike, it is not always completely consistent.

Foreign concepts of ecological seadikes are also different. The Netherlands has proposed the ecological protection concept of "returning to nature with nature", such as the Suder Sea Project and the Delta Project. In the above projects, an open tide gate was built and an ecological buffer zone was constructed with the wetland as the core, realizing the combination of ecological environmental protection and water conservancy engineering construction. Not only does it solve the coastal erosion problem faced by the Netherlands, but it also stops the salinization of polder fields. The United States follows the "build with nature" protection concept and has applied it in the construction of ecological seadikes. A representative of this is the Sack Coastal Protection Project in Maine. In this project, ecological approaches such as beach nourishment, dune restoration, and marsh vegetation planting have been integrated and realized, which not only improved the seawall's marine disaster defense capability, but also provided a large amount of habitat for marine organisms. Australia has adopted the concept of "harmony between human beings and nature" and has made ecological changes in the marine protection of New South Wales. Its main means include seadike openings, artificial stacks, stone throwing protection, and the combination of dunes and vegetation, etc., building a diverse ecological protection barrier. In terms of results, the above measures can significantly reduce the impact of anthropogenic factors on the ecological seawall structure and provide the necessary spatial place for the habitat survival of marine organisms.

Summarizing the practical experience of ecological seawall construction and related research progress at domestic and international level, ecological seawall is a coastal protection building with the ability to meet the standard tide protection and disaster prevention, which is unified with the natural coastal ecological function.

A typical ecological seadike concept is shown in Fig. 4.6.1.

Fig. 4.6.1 A typical ecological seadike concept

4.6.2 Ecological seadike composition and construction principles

The spatial composition of the ecological seadike basically consists of 3 parts: the wetland zone in front of the dike, the structural zone of the dike, and the buffer zone behind the dike, as in Fig. 4.6.2.

1. The wetland zone in front of the dike

The wetland zone in front of the dike should have a more complete intertidal topography structure, a stable and rich biological community, and a natural and harmonious wetland landscape, providing the first barrier for the seawall to eliminate waves and prevent washouts.

Chapter 4 Seadike Engineering

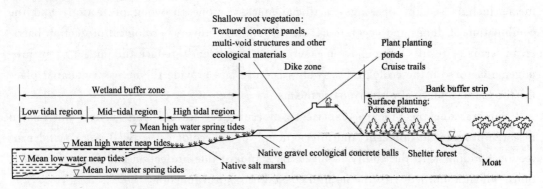

Fig. 4.6.2 Schematic diagram of ecological seadike structure section

2. The structural zone of the dike

The dike structure should be smoothly connected to the wetland zone in front of the dike and the buffer zone behind the dike, with a natural dike or near-natural dike to meet the standard of physical energy exchange capacity, and is the main barrier to prevent the tides and disasters.

3. The buffer zone behind the dike

The buffer zone behind the dike is the transition zone of the coastal ecosystem linkage, the coastal zone dissection and important protection circle, with the ability to mitigate the impact of disasters and reduce the risk of seawall destabilization.

Ecological seawall construction should meet the following principles.

(1) Safety standards compliance. The primary function of the ecological seawall is to prevent tides and disasters. According to the economic and social characteristics of the protected area, its ability of mitigation tides and disasters, formed by the combination of its shape and structural stability, should meet the corresponding specification requirements or design standards.

(2) Ecological suitability. Construct the ecological seawalls according to the natural ecological characteristics of the coast and the habits of important species such as breeding, habitat and migration. The zone composition and structural materials of ecological seadikes are suitable to protect and develop the original ecosystem of the nearshore area.

(3) Functionality enhancement. The structure and function of ecological seawalls will yield a number of services, including maintenance of biodiversity, formation and protection of intertidal substrates, habitat creation, cleaning of incoming pollution, landscaping, disaster prevention and mitigation, and recreation.

Exercise

Q1 Fill in the blanks: Distinguished by the shape of the waterfront, the types of seadikes are ( ), ( ) and ( ).

Exercise

Q2　Fill in the blanks: The main focuses on the design of each dike type including (　　), (　　), (　　), foundation strength, uneven subsidence, wave energy dissipation.

Q3　Short answer: What are the principles of seawall type selection?

Q4　Short answer: How do we choose a seawall type?

Q5　Multiple choices: Seawall armour materials include (　　).

A. Dry block or bar stone   B. Slurry block
C. Stone throwing   D. Precast concrete panels
E. Cast-in-place integral concrete   F. Asphalt concrete
G. Artificial blocks   H. Cemented soil and turf armour

Q6　Fill in the blanks: Sloping seadike footing rock prism role includes (　　), (　　).

Q7　Short answer: What is the role of seadike backfilter layer?

Q8　Judgment question: The top elevation of the seawall is the top elevation of the seawall when it was just completed.

Q9　Multiple choices: Factors affecting the determination of seawall top elevation mainly include (　　).

A. Recurrence period of tidal level and waves in seawall defense criteria

B. Wave train cumulative frequency

C. Local meteorological and hydrological conditions

D. Structural characteristics of the seadike

Q10　Fill in the blanks: The seawall top elevation calculation consists of 3 parts: (　　), (　　) and (　　).

Q11　Multiple choices: Factors affecting the top width of the dike include (　　).

A. Self-stabilization

B. Foundation stability

C. Wave and seepage prevention requirements

D. Construction and flood prevention and rescue requirements

Q12　Short answer: What are the factors affecting the slope of the side slope of a sloping dike? What is the procedure for determining the dike side slope during the design process?

Q13　Fill in the blanks: The top width of the dike should mainly meet (　　) and (　　).

Q14　Short answer: What is the crest of the dike designed to meet the functional needs of traffic?

Q15　Short answer: What are the functions and basic requirements of armour?

Q16　Short answer: What are the features of dry block armour?

Q17　Short answer: What are the features of slurry masonry or grouted block stone armour?

Q18 Short answer: What are the features of seawall rock throwing armour?

Q19 Short answer: What are the features of concrete slab surface protection?

Q20 Short answer: What are the characteristics of artificial block armour for sloping dikes?

Q21 Short answer: Give some basic experience on the design of the backfilter layer.

Q22 Fill in the blanks: The main role of armour footings is (   ). Its structure type includes (   ), (   ) and (   ).

Q23 Short answer: The main role of steep-wall seawall protective wall?

Q24 Multiple choices: Main precursors of seawall or armour instability include (   ).

A. Dramatic increase in subsidence

B. Rising slope corners

C. Longitudinal cracks in the dike body

D. Increased void water pressure and seepage at slope corners

Q25 Fill in the blanks: Seadike stability calculations include (   ) and (   ).

Q26 Fill in the blanks: Main methods of slip stability calculation (   ), (   ), (   ), $\varphi=0$ method, (   ) and Sliding surface simplification calculation method, etc.

Q27 Fill in the blanks: Foundation subsidence includes (   ) and (   ).

Q28 Multiple choices: Seadike foundation reinforcement method include (   ).

A. Displacement sand layer method—also called dredging method

B. Suppression layer method

C. Drainage sand bedding method

D. Reinforced geotextile laying

Chapter 5 Protection Engineering

5.1 Introduction

5.1.1 Functions and basic facilities of coastal protection engineering

Protection engineering is an important part of coastal engineering. For the purpose of coastal protection (or shore protection), different types of coastal engineering or other coastal protection measures can be constructed. The functions of coastal protection engineering mainly includes preventing coastal erosion, stabilizing the shoreline and protecting the backshore part of the waterfront or filling the land area.

The basic structures of coastal protection include groins, detached breakwaters, revetments, seawalls and other coastal structures, as well as measures such as artificial beach nourishment and ecological coast protection. A short overview of the features and functions of these basic protection engineering is given below.

1. Groins

A groin is a coastal structure arranged roughly perpendicular to the shoreline. The length of a groin is about 0.4 - 0.6 times of the average width of the local surf zone from shoreline to offshore. The groin will block the part of longshore sediment transport in the surf zone. A single groin structure will intercept longshore sediment transport, resulting in accretion on the upstream side and erosion on the downstream side.

In order to protect the shoreline, it is necessary to build several groins along the shoreline to form a groin group. Generally, the spacing of a groin is about 1 - 3 times of the length of a groin. Longshore sediment transport intercepted by the groin group will be deposited on the upstream side of the groin group and on the beach between the groins, thus preventing the erosion of this section of coast. When adopting the groin group as a coastal protection engineering, attention should be paid to preventing the erosion of the downstream side beach of the groin group.

2. Detached breakwaters

The construction of a breakwater that is roughly parallel to the coast at a certain distance from the shoreline is called offshore breakwater in marine engineering and detached breakwater in coastal protection. Usually, offshore breakwaters are built in relatively deep waters to guarantee sufficient harbor water area on their back side, while detached breakwaters are built in relatively shallow waters closer to the shoreline to provide effective protection for the beach.

The wave energy is reduced in the area behind the detached breakwater; hence, the detached breakwater can effectively protect the beach from the erosion of the waves. Within the range between the detached breakwater and the shoreline, the longshore sediment transport rate will also be reduced, contributing to the deposition of sediment input from the upstream side.

Detached breakwater can be arranged as a single dike, or an interrupted form which is a segmented detached breakwater with an entrance between every two short dikes. The downstream side of the detached breakwater also suffers from beach erosion because longshore sediment transport is intercepted on the upstream side. Practice shows that the impact of the segmented detached breakwater on the downstream beach is less than that of the single dike with the same total length.

3. Revetments and seawalls

A revetment or seawall is a structure built on the higher part of the beach to demarcate the land area of the beach from the sea, and its course is generally parallel to the shoreline. In coastal protection engineering, there is no clear definition to distinguish between revetments and seawalls. Generally, buildings mainly retaining soil on the boundary of sea and land can be called revetments. The top elevation of the revetment (excluding the height of the wave wall at the top of the revetment) is commonly the same or close to the elevation of the land behind it. During storm surges and large waves, buildings whose main purpose is to protect land areas and structures from seawater flooding and wave damage can be called seawalls. The top elevation of the seawall is often higher than the elevation of the land behind it. A revetment or seawall only protects the coastal land behind it and does not prevent or reduce erosion on the beach in front of it. In the case of a vertical revetment, the erosion of the beach in front of it will be intensified because of its reflection on the waves.

4. Artificial beach nourishment

The most natural response to beach erosion is to collect suitable sand from the sea or onshore to replenish the eroded shoreline. Beach nourishment has been proven to be a cost-effective measure and has less impact on the downstream shore than other protection engineering.

The beach must be replenished with sand every few years because the sand artificially filled into the beach will still be washed away by various marine environmental conditions, especially by waves.

5. Ecological coast protection

The ecological coast protection project refers to a measure of planting vegetation to resist the impact of external dynamic factors such as tides and waves in coastal protection engineering in a low-energy environment, thus the coastal embankment can be protected from erosion.

5.1.2 Causes of coastal erosion

The essence and extent of coastal erosion varies from region to region. In order to determine reasonable coastal protection engineering, a systematic and in-depth study should be conducted into the marine environmental conditions and shore conditions of the region. An important task of the study is to clarify the causes of local coastal erosion.

The most severe coastal erosion occurs during a severe storm surge. However, the actual causes of coastal erosion are multifaceted. Some of these causes are induced by natural factors and some by human-induced factors. An overview of the causes of coastal erosion is as follows.

1. Sea-level rise

A slow rise in mean sea level relative to land has been occurred in many regions of the world. The shore profile will gradually adjust to the relatively high mean sea level, resulting in slow long-term erosion of the shoreline.

2. Variation of sediment supply within the inshore area

In arid areas, the reduction of river runoff will result in less sediment supply within the inshore area.

3. Storm waves

Steep waves during a storm will transport sediment from the beach toward the offshore and create a sand bar or shoal. In contrast, long-period waves after a storm will transport sediment to shore and restore part of the beach. However, in most cases, some beach sediments will remain permanently in relatively deep waters.

4. Scouring of the upper beach by huge waves

During storm surges, huge waves can wash the upper beach, which will cause sediment to transport and deposit toward the backshore.

5. Sediment transport by wind

The gale can blow loose, fine-grained sand from the beach, causing coastal erosion. On the backshore, eolian dunes of considerable size are formed.

6. Longshore sediment transport

The alongshore current formed by the wave breaking through the oblique transmission of the wave to the beach will cause longshore sediment transport. If the sediment carrying capacity of the alongshore current exceeds the sediment supply of the upstream section, the beach will erode.

7. Land subsidence due to exploitation of underground resources

The exploitation of underground resources such as oil, gas, water, or coal in the coastal zone can cause subsidence of land and beaches. The adjustment of the shore profile will cause the shoreline to erode slowly for a long time.

8. Interception of longshore sediment transport

One of the major causes of coastal erosion is the interception of all or part of longshore sediment transport due to human production activities. For example, the regulation of a tidal inlet, whether it is a jetty or a channel, will create a barrier to longshore sediment transport. This will cause erosion of the downstream side beach.

9. Decrease in sediment supply within inshore areas

In some areas, riverine sediment supply is the main source of sediment within inshore areas. The construction of floodgates on rivers can reduce the supply of sand to the nearshore zone, thus causing coastal erosion.

10. Increase of local wave energy

The construction of coastal structures within the beach area, such as vertical revetment, will reduce the energy lost by the original wave propagation on the beach. Waves reflected by vertical structures can cause erosion of the beach in front of them.

11. Changes of natural coastal protection

The excavation of sand bars or shoals along the shore will increase the wave energy acting on the beach, thus causing shore erosion. Moreover, the removal of sand dunes on the upper beach can also further lead to the scouring of the upper beach by huge waves.

12. Beach sand mining

Beach sand mining will disrupt the balance of sediment supply and demand in the nearshore zone, thus causing further erosion.

Among the above-mentioned causes of coastal erosion, the first six are natural causes and the last six are human-induced causes.

5.1.3 Selection of coastal protection engineering

For the four basic types of coastal protection buildings such as groins, detached breakwaters, revetments and artificial beach nourishment, the Permanent International Association of Navigation Congresses (PIANC) gives a comparative table of applicability which the shoreline sediment is dominated by vertical movement (longshore sediment transport) and transverse movement (sediment transport to and from the shore) (Table 5.1.1). Moreover, longshore sediment transport is further divided into four cases, such as strong, weak, one-way and two-way longshore sediment transport.

Table 5.1.1 Selection of coastal protection engineering

Coastal protection engineering				Groins	Detached breakwaters	Revetments	Beach nourishment
Sediment movement	Longshore sediment transport	Strong longshore sediment transport	One-way	A	D	E	D
			Two-way	D	C	E	D
		Weak longshore sediment transport	One-way	A	D	D	B
			Two-way	D	B	D	C
	Transverse sediment transport			D	B	C	A

Note: A—most appropriate; B—appropriate; C—minor; D—poor; E—inappropriate.

Based on the introduction of the characteristics of various coastal protection engineering in Section I and the comparison of applicability in Table 5.1.1, the following points can be summarized for reference in the selection of practical engineering solutions:

(1) The groin group is suitable to be constructed in the case where the sediment is mainly transported along the coast and the dominant direction of sediment transport is obvious (it is mainly one-way sediment transport). However, the groin group is basically ineffective for coastal protection dominated by transverse sediment transport.

(2) Detached breakwaters are mainly suitable for coasts dominated by transverse sediment movement and are also effective for intercepting longshore sediment transport.

(3) Artificial beach nourishment is mainly suitable for transverse sediment movement, but not suitable for the situation of strong longshore sediment transport.

(4) For the purpose of coastal protection, revetments are often used in combination with groin systems or artificial beach nourishment rather than individually.

The design wave criteria given in Chapter 2 is suitable for marine structures such as groins, detached breakwaters, and revetments.

5.2 Revetments

5.2.1 Cross-section shape of revetments

The difference between a revetment as a coastal protection project and a revetment generally built in a sheltered harbor or on a riparian lies mainly in the fact that the former needs to consider the wave action, including the wave force acting on the revetment structure, the crested water body over the top of the revetment and the local scouring of the bottom in front of the revetment by the waves, etc.

The revetment is usually built on the relatively high part of the beach. As the boundary between the seaside land and the sea, one of its functions is to prevent the beach from receding to its rear side, and the other is to protect the backshore of its rear shore or filled land area.

The cross-section shape of the revetment is one of the main elements to be selected and determined in the design. It can be roughly divided into vertical or steep, sloping, concave curve and stepped types, etc. (Fig. 5.2.1).

When the near-sea side of the revetment is a vertical surface or close to a vertical surface (steep wall), the advantage is that it can be used as a quay wall to dock boats when there is no wind and waves. However, the disadvantage is that the wave reflection is large and the erosion in front of the wall is serious. The depth of the local scour pit in front of the vertical wall or steep wall built on the bank slope is approximately equal to the maximum significant wave height in front of the wall, that is H_S. When the wave has broken before reaching the wall, the depth of the scour pit can take the limiting broken wave

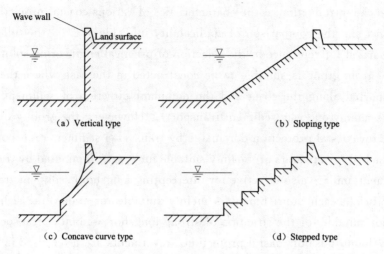

Fig. 5.2.1 The cross-section shape of revetments

height H_{Sb} determined by the water depth at the toe of the wall.

The slope revetment is mainly composed of a riprap structure, which is conducive to wave energy dissipation and absorption, so the wave run-up on the slope and overtopping on the top of the revetment are both small. When the slope revetment adopts concrete block with good wave suppression performance, such as tetrapod or dolosse, the wave run-up and overtopping can be further reduced. The erosion of the beach in front of the slope revetment is also smaller than that of the vertical type.

Throwing riprap prisms and artificial blocks in front of the vertical wall and making their top surface flush with the top surface of the wall can effectively weaken the effect of waves on the wall, reduce the wave overtopping and decrease the scouring in front of the wall.

The revetment with a concave curve shape not only has a beautiful appearance, but also helps to reduce the wave overtopping. When there is a road behind the seaside tourist area or the revetment, this type of revetment is often the main solution to be considered.

The stepped revetment is suitable for construction in seaside tourist areas. Because steps can be used to easily reach the beach from the rear land at low tide. Moreover, steps also help to reduce the scouring caused by the falling waves.

5.2.2 Estimation of beach erosion profile before revetments

In the design of revetments, it is important to distinguish the difference between local scouring before revetments and long-term erosion of the beach before revetments on the erosive coast. Local scouring is a short-term change, which can be basically formed during a large wind wave process.

The exact impact of revetments on the erosion process on a coast that is continuously eroded year by year is not yet fully understood. A simplified method for predicting shore-

5.2 Revetments

line change before revetments is given in the US *Shore Protection Manual*. For the long-term effects of the revetments, it is first assumed that the buildings do not have a mitigating effect on the erosion of the beach on their front side and this erosion is continuous and uninterrupted.

The solid line in Fig. 5.2.2 shows the existing shore profile with a planned revetment at point A. According to the previous information, we can know the width of the receding beach per year. The analysis of the profile of the eroded beach shows that the beach beyond about 10 m water depth will remain basically unchanged, the turning point E of the beach will be maintained at the same elevation during the beach change and the beach slope on the shore side of the turning point can be considered to be basically the same. Based on the above conditions, the final erosion depth of the beach at the location of the revetment axis can be estimated as follows.

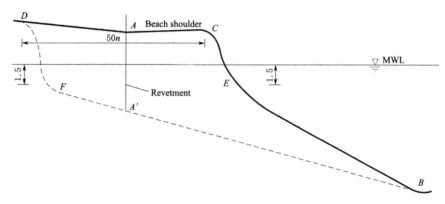

Fig. 5.2.2 The beach erosion profile before revetments

(1) In Fig. 5.2.2, it is assumed that point B is the intersection of the beach and the 10 m water depth line, point E at a depth of about 1.5 m is the turning point of the beach slope and point C is the front edge of the berm. If the service life of the revetment built at point A is required to be 50 years and the annual erosion width of the berm is known to be n, then the erosion distance of the berm will be $50n$ when there is no revetment, that is, the berm will retreat to point D in the figure.

(2) Draw a new beach line DF which is parallel to CE from point D downward in which point F is the turning point of the beach slope after erosion and its depth is the same as that of point E. Then connect point F with point B. The dashed line DFB on the figure will represent the approximate erosion profile of the beach after 50 years in the absence of a revetment. The shoreline at the location of the revetment will be eroded to point A' after 50 years, from which the depth of erosion at that location can be obtained.

As can be seen from the figure, the construction of a revetment on an erosive shore generally requires a certain distance back from the existing shoreline, otherwise the depth of erosion at the revetment will be too deep after a certain number of years. However, if the revetment is constructed at a certain distance back from the existing shoreline, a considerable amount of excavation is required during construction unless a structure such as sheet pile or diaphragm wall is used. On the flip side, although the back side of the line AA' is protected from erosion by the revetment within the length of the revetment, the line AD must be constructed with wing walls at both ends of the revetment length because the beach profile is eroded to $DFA'B$. Otherwise, the revetment ends will be subjected to lateral scouring and the structural safety will be compromised. The erosion of the beach will also intensify on the downstream side of the revetment.

Through the above analysis, revetments should not be used alone for the purpose of coastal protection. Of course, revetments are still widely used coastal structure when combined with a system of groins or artificial beach nourishment, or when built on a non-erosive shore to protect filled land area of its backside.

5.2.3 Permissible overtopping discharge of revetments

Generally, the revetment that protects the coastal land is not required that the waves will not overtake the revetment at all during high tide and huge waves, but it is required to control a certain crest water body that crosses the top of the revetment (including the height of the break wall), that is, permissible overtopping discharge. Allowing overtopping, on one hand is for economic reasons, on the other hand is for the limited length of revetment. When the entire coastal section is submerged by seawater, the revetment wall will not be able to prevent the seawater from invading the land behind it from both sides.

The method of calculating the wave overtopping of straight wall and slope revetments still uses the formula for the overtopping of straight wall and slope buildings, where the criteria for the permissible overtopping discharge need to be given. Based on the results of the actual engineering survey, the standards for the permissible overtopping discharge are divided into two types: one is mainly to ensure structural safety, the other is to ensure land use. In order to ensure that the revetment structure and the ground behind it are not damaged by overtopping waves, the permissible overtopping discharge $[q] = 0.2 \text{ m}^3/(\text{s} \cdot \text{m})$ when the ground has masonry or concrete pavement and $[q] = 0.05 \text{ m}^3/(\text{s} \cdot \text{m})$ when there is no pavement on the ground. To ensure the land use behind the revetments, it is generally advisable to take $[q] = 0.01 \text{ m}^3/(\text{s} \cdot \text{m})$. If there are no important industrial facilities and buildings on the land, and there are suitable drainage measures, the permissible overtopping discharge is allowed to increase appropriately.

Even if there is a standard for permissible overtopping discharge and the calculation method of overtopping discharge, it is still necessary to clarify what design tide and design wave are used to calculate wave overtopping. According to practical experience, the design tide and design wave standard can be specified as follows: the design tide adopts the extreme high tidal level and design high tidal level in the *Code of Hydrology for Harbour and Waterway* (JTS 145—2015), respectively. The extreme high tidal level is the 50-

year high tidal level, which is mainly used to account for the shore protection structure and its background from being damaged by the overtopping waves. The corresponding design wave recurrence period is also 50 years, which can be reduced to 25 years when the requirement is relatively low. The design tide and design wave standard are equivalent to the extreme high tidal level combination condition in the lasting state when the building is designed according to the probabilistic limit state design, or the calibration combination condition when the building is designed according to the safety coefficient method.

The design high tidal level is 10% of the cumulative frequency of high tide, which is mainly used to account for the use of the land area behind the revetments. The corresponding design wave recurrence period can be determined according to the use requirements and land drainage facilities, etc. The high value is 50 years and the low value is 10 years.

5.2.4 Structure types of revetments

The following will introduce the revetment structure types in coastal protection engineering according to the classification of the revetment section shape.

For vertical or steep slope revetment, structures such as precast concrete squares or precast reinforced concrete caissons commonly used in harbor engineering can be used. However, if the local tidal range is large and the waves are small during the construction period, the stone masonry steep wall shown in Fig. 5.2.3 can be used, which is more economical. Because of the lateral pressure of the fill or rockfill behind the wall, the vertical revetment is usually designed for the most unfavorable situation when the wave trough appears in front of the wall. As shown in Fig. 5.2.3, the design wave height of the revetment section is $H_{1\%} = 2.3$ m.

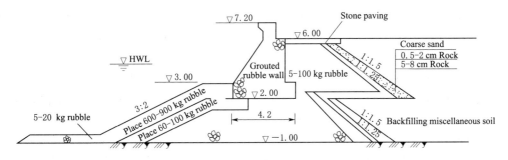

Fig. 5.2.3 The thick foundation bed stone masonry steep slope revetment
(unit in elevation: m, scale: m)

The revetments built in the seaside tourist scenic areas need to take into account the requirements of the landscape. Fig. 5.2.4 shows the cross-section view of a wave dissipation block vertical revetment built in the seaside area of Xiamen. The main structure of the vertical revetment adopts a kind of wave dissipation block called "Tender Blocks". The block is mainly composed of a horizontal top plate with open holes and a vertical structure with an arc-shaped front end. When constructing the revetment wall with tender blocks, the upper and lower layers of the blocks are staggered, which not only form the wave dissipation channel, but also form a beautiful appearance in which the arc surface and the hole alternate up and down, left and right.

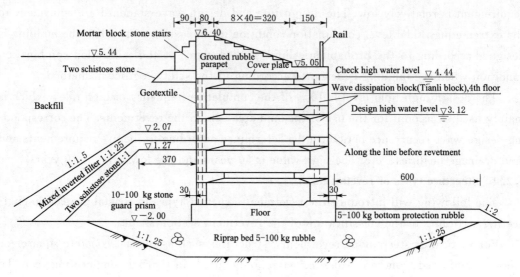

Fig. 5.2.4 The cross-section of the wave dissipation block revetment
(unit in elevation: m, scale: m)

In Fig. 5.2.4, from the -2.0 m seafloor elevation upward, the bottom layer of the wall of the wave dissipation block revetment is the reinforced concrete floor, and there are four layers of "Tender Blocks" wave dissipation block on the floor. Each block weighs about 25 t and the reinforcement ratio is about 65 kg/m^3. The reinforced concrete cover plate is placed on the top layer of the block and then the stepped-type stone masonry breast wall is built. The stepped-type breast wall is convenient for visitors to enjoy the sea view at the outside of the breast wall when there is no wind and waves.

The revetment wall which is composed of wave dissipation blocks has the effect of reducing the reflected waves in front of the wall, and its wave reflection coefficient is about 0.4 - 0.5. Since the wave reflection in front of the wall is small, the wave crest and wave trough forces on the wall are also small. In addition, the wave overtopping discharge at the top of the wall and the scouring of the waves on the sea floor in front of the wall are relatively small.

Slope revetment is the most widely used revetment structure type, and its surface type is similar to that of slope breakwater groin. Above the construction tidal level, paved rock blocks or stone masonry armour layers can be used. Below the construction tidal level, throwing or laying stone armour layers can be used. When the waves are large, reinforced concrete fence plates or concrete four-legged hollow cubes armour layers are often used in China. In other countries, asphalt concrete is also used as an armour layer. Fig. 5.2.5 shows the cross-section of a fence plate slope revetment.

5.2 Revetments

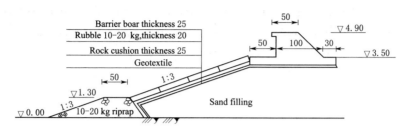

Fig. 5.2.5 The fence plate armour sloping revetment
(unit in elevation: m, scale: cm)

Fig. 5.2.6 shows the cross-section of a slope revetment which protects nuclear power plants. On the one hand, according to the requirements of nuclear safety, it is necessary to adopt relatively high design standards. The DBF tidal level in the figure is the "basis flood level" in the design of a nuclear power plant, which is equivalent to the check high tidal level under extreme conditions; and in this case, only the magnitude of meter level of the wave overtopping discharge is allowed at the top of the revetment wall. On the other hand, the factory also requires that the top of the wave wall should not be too high, which for the needs of viewing the sea behind the wave wall. To this end, two layers of dolosse are used for the slope armour structure. The wave climbing height on two layers of twisted I-shaped blocks can be about 24% lower than that on a single layer of hook-connected blocks or twisted W-shaped blocks, and the wave overtopping discharge is also smaller when the two-layer dolosse is used. In other words, when controlling the same amount of overtopping, the use of dolosse armour layers can reduce the height of the wave wall. In addition, the submarine surface of the revetment is relatively thick sludge, and a fairly low slope of 1:2 should be used for the outer slope of the revetment in combination with the need of foundation stability, as well as a secondary wide platform is set in the middle and top of the revetment to reduce wave run-up and wave overtopping. In order not to block people's view of the sea behind the wave wall, a gentle slope of 1:10 toward the sea is adopted at the top of the dolosse platform at the top of the wall.

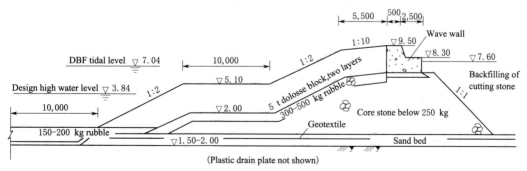

Fig. 5.2.6 The dolosse armour sloping revetment
(unit in elevation: m, scale: cm)

169

Fig. 5.2.7 and Fig. 5.2.8 show the concave curve revetment and the stepped revetment, respectively, both of which are built at the relatively high part of the beach. Moreover, the curved concrete wall and the reinforced concrete step were constructed at low tide. Due to the large scouring deformation of the shoreline, sheet-pile walls were placed under the revetment walls.

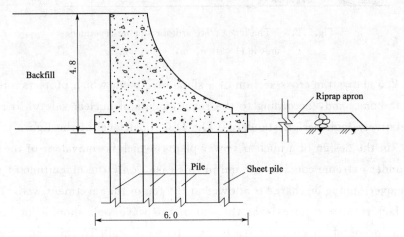

Fig. 5.2.7 The cross-section of the concave curve revetment (unit: m)

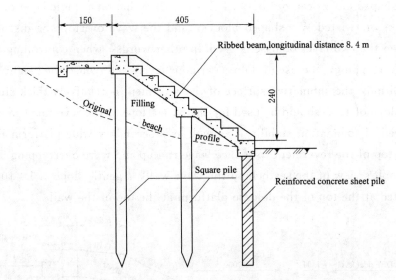

Fig. 5.2.8 The cross-section of the stepped revetment (unit: cm)

5.3 Groynes

5.3.1 Two different scour and deposition morphology

Groynes are one of the most widely used coastal structures, but their interaction the-

5.3 Groynes

ory is still not fully understood. As shown in Fig. 5.3.1 (a), groynes have the effect of intercepting longshore sediment transport, causing siltation on its upstream side and scouring on its downstream side, which mainly occurs on sandy coasts. In the case of a groyne structure built on a muddy coast, the water flow formed on the upstream side of the groyne due to the breaking of the obliquely incident wave will scour the beach which composed of the fine sediment, while the wave cover area on the downstream side of the groyne will become the deposition area of suspended sand, as shown in Fig. 5.3.1 (b). It is possible to form different scouring and deposition morphology on the upstream and downstream sides of the groyne because of different sizes of sediment particles. Fig. 5.3.1 shows the typical beach scour and deposition comparison graphics between the upstream and downstream sides of the groyne on a sandy coast and a muddy coast. This completely different deposition and scour situation due to different environmental conditions should be given great attention when selecting the scheme of coastal protection engineering.

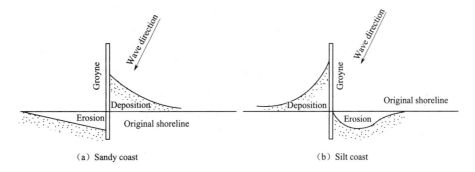

Fig. 5.3.1 The scouring and deposition graphics of the upstream and downstream sides of groynes

The discussion in this chapter, except for note, refers mainly to the case of sandy coasts.

5.3.2 Types of groynes and their length and height

Groynes can be divided into basically impermeable type and sand-pass type based on the sand permeability. According to the crest elevation, they can be divided into the high groyne and the low groyne. According to the length, they can be divided into the long groyne and the short groyne. Most groynes are fixed structures, but a few are movable structures. The purpose of a sand-pass groyne is to allow some of the intercepted longshore sediment transport to pass through the groyne to slow down the scouring behind the groyne. Most groynes, including riprap structures, are essentially impermeable type. The purpose of sand permeability can also be achieved by lowering the dam crest elevation.

The main type of movable groynes is a structure with long precast slabs placed on spacer piles. The slabs can be increased or decreased to adjust the top elevation of the groyne so that some of the sediment can be transported to the back of the groyne. However, when this structure is displaced, it will be difficult to add and subtract slabs.

The length of groynes can range from about 30 m to several hundred meters, which mainly depends on the width of the surf zone of local general storm waves and the ratio of the planned interception of longshore sediment transport. The latter is related to the degree of protection required by the beach. Since most of longshore sediment transport occurs within the coastal surf zone, it is not necessary to extend the head of the groyne beyond the breaker line. Physical model tests have shown that the length of the groyne should be 0.4 - 0.6 times of the width of the surf zone. If the groyne extends the shoreline too long, the scouring on the downstream side of the groyne or the groyne group will be relatively serious.

In the actual engineering design, there is no clear definition about how to determine the so-called "the width of general storm wave surf zone" or "the width of average surf zone" in the general coastal engineering literature. Based on experience, the following tide and wave criteria are recommended to determine the "the width of general storm wave surf zone". The tidal level, which depends on the degree of shore protection requirement, can be selected from the average high tide level or the tidal level of 10% guaranteed rate of high tide. The latter is the design high tidal level of the harbor project, which is roughly equivalent to the average high tidal level of the spring tide. The recurrence period of the design wave can be 2 - 5 years, and its characteristic value in the wave group is set as the significant wave height H, and the average period T. According to the method of breaking wave function, the water depth of the breaking wave is determined. Therefore, the width of the surf zone can be obtained.

Along the length of the groyne, it can usually be divided into three sections: the first section is the shore side section connected to the bank; the second section is the middle section with a longitudinal slope at the top of the groyne; and the third section is the outer section including the head of the groyne (Fig. 5.3.2).

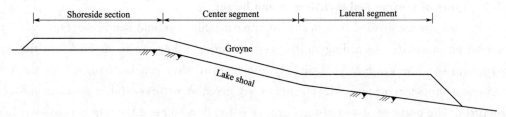

Fig. 5.3.2 The longitudinal section of the groyne

The starting point of the shore side section is usually located at a certain distance from the front edge of the berm (beach platform) of the beach section to the shore side, so as to prevent the root of the groyne from being washed out.

The standards for high and low top elevation of the shore side section are given in the USA *Shore Protection Manual*. The low standard is the highest high tidal level plus the usual wave run-up on the beach. Moreover, it is often necessary to add another 0.3 m of super-elevation to reduce sand penetration in the case of the riprap groyne; the high standard is applicable to high groynes that want to intercept all the upstream sediment, using the highest tidal level plus the storm wave run-up on the beach. Since the highest tidal level is usually caused by storm surges with a long recurrence period, the above criteria seem to be relatively high, and there is no exact definition of the so-called "normal wave" and "storm wave". In general, for the top elevation of the shore side of the groyne, it can be applied 10% of the cumulative frequency of high tidal or averaged high tidal level of the spring tide, then adding 0.6 - 0.7 times of the limited breaking value of the significant wave height H_S with the recurrence period of 2 - 5 years, i.e. $(0.6 - 0.7)H_{Sb}$. If the degree of protection requirement for the beach is low, the top elevation of the shore side of the groyne should be taken as the above tidal level plus 1.0 - 1.5 m super-elevation.

The crest elevation of the middle section is gradual and the longitudinal slope of the crest should be roughly parallel to the slope of the beach surface of the section. The crest elevation of the outer section should be higher than the beach surface or seafloor elevation after the sediment has been intercepted on its upstream side. The top elevation of the groyne also depends on the construction method of the groyne. If using the method of land proceeding, the top elevation should not be too low to be affected by the tide and waves during the construction period.

5.3.3 Plane layout of the groyne group

The distance between two adjacent groynes in the groyne group is related to the length of the groynes and the shoreline orientation after erosion and deposition changes. In the case of one-way longshore sediment transport, the shoreline orientation between the two groynes will be approximately perpendicular to the representative wave direction (dominant wave direction or synthetic wave direction) of the shallow water waves after refraction. In Fig. 5.3.3, two groynes, groyne 1 and groyne 2, are shown and the longshore sediment transport direction is from left to right. When longshore sediment transport is blocked by groyne 1 and gradually accumulates on its upstream side, the shoreline sediment is transported from left to right due to the effect of the longshore current without upstream sediment supply between the two groynes, thus causing the left part of the shoreline to be eroded and the right part to be silted up. When the upstream side of groyne 1 is silted fully, the shoreline between groyne 1 and groyne 2 will have the maximum change. The line *ab* in Fig. 5.3.3 indicates the front edge line of the berm at this time, and the area of the eroded part A is approximately equal to the area of the siltation part B. Groyne should ensure that there is sufficient length after the deepest erosion position *a* of the shoreline on its downstream side to prevent the root part of the groin from being eroded. After the upstream side of groyne 1 has been silted, the upstream sediment will cross groyne 1 and enter the water area between groyne 1 and groyne 2 (dam field) and be partially deposited. The relatively stable shoreline is shown as line *cd* in the figure. Then, the upstream sediment will cross groyne 2 and be transported to its downstream side.

Chapter 5 Protection Engineering

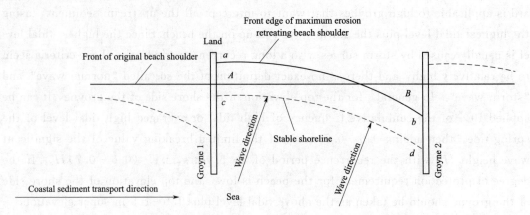

Fig. 5.3.3 The plane layout of the groyne group

When the direction and intensity of incoming waves have obvious changes, the shoreline change between the groynes will be more complex.

The general rule recommended in the USA *Shore Protection Manual* for determining the distance of groynes is that the distance should be equal to 2 – 3 times of the total length of the groyne (including the portion extending from the leading edge of the berm into the shore side).

On the downstream side of the groyne group, shoreline erosion on the downstream side will occur because the sediment carrying capacity of the longshore current is greater than the upstream sediment supply, unless there is a significant change in the shoreline orientation that makes the longshore component of wave energy decrease significantly. An effective solution to shoreline erosion on the downstream side is to artificially fill the sand on the upstream side of the groyne group and between the groynes. As a result, longshore sediment transport is basically not intercepted, and can be transported to the downstream beach through the head of the groyne group.

When the intersection angle between the dominant wave direction and the shoreline is large, a T-shaped or L-shaped groin head has been designed to enhance the protection of the beach. However, the wave scouring to the T-shaped or L-shaped groyne head is greater than that to a conventional linear groyne.

In the tidal estuary, groynes are generally arranged roughly orthogonal to the direction of current. In order to reduce scouring of the groyne head, the axis of the groyne can be slightly oriented to the direction of the tidal current with relatively low velocity. In order to accelerate siltation between the groynes, the axis of the groyne can be slightly oriented to the direction of the tidal current with relatively high sediment concentration.

5.3.4 Layout and scale of groynes for beach protection and siltation promotion

Generally speaking, groynes have the function of promoting siltation and protecting the beach. The beach-protecting groynes in this section refer to the groyne built in front of the seawall to prevent the foot of the seawall from being washed away to protect the seawall foundation, while the siltation-promoting groynes refer to the groyne built on muddy beach with gentle slopes to promote the siltation of the beach surface to meet the requirements of the reclamation development.

The beach-protecting groynes are generally short groynes with a length of about 50 – 150 m, and about 100 m is the most suitable. The siltation-promoting groins are usually long groynes with a length of 1,000 – 2,000 m.

The distance between two adjacent groynes in a short groyne group is generally 2 – 4 times of the length of the groyne, of which 2 – 3 times for concave shores and 3 – 4 times for straight shores. The distance between the two groynes of the siltation-promoting long groyne group is relatively small, which is generally not more than 2 times of the length of the groyne.

The connection line of each groyne head in the groyne group should be roughly parallel to the direction of the tidal current. When the groyne adopts a hook form (equivalent to an L-shaped groyne head), the hook head should be oriented to the dominant sediment direction.

For siltation-promoting projects, long groynes are often used in combination with longitudinal groynes (detached breakwaters) in order to increase the effect of siltation promotion.

The crest elevation of a siltation-promoting groyne can be set roughly at the mean high tidal level, in which case the siltation between the groynes can reach the elevation between the mean tidal level and the mean high tidal level. When the groyne crest elevation is higher than the mean high tidal level, the siltation does not increase significantly. In the case of a beach-protecting groyne, since its main purpose is to protect the beach from erosion, the groyne crest elevation can be relatively low to reduce the project cost.

A longitudinal slope from the shore to the sea can be set along the top surface of the groyne. For short groynes, the gradient of the longitudinal slope can be about 1‰; for long groynes, the longitudinal slope is smaller.

5.3.5 Structural types of groynes

The sloping riprap structure is the most common type of groyne. In the higher part of the beach, stone masonry armour can be used; in the lower part of the beach, block stone armour or lay stone armour can be used instead of paved rock blocks which are difficult to work. Fig. 5.3.4 shows a typical cross-sectional view of the groyne with two layers of paved rock blocks and block stone. If the local waves are large and the supply of large blocks is difficult, reinforced concrete fence plates or concrete four-legged hollow cubes can be used as armour layers in relatively deep sections. Along the length of the groyne, the height of the riprap prisms can be adjusted to suit the scale requirements for placing fence plates or four-legged hollow cubes on the slope due to the variation of the height of the groyne and the length of the slope surface. Fig. 5.3.5 shows the cross-sectional view of groynes with fence plate armour layers. When the groyne adopts the fence plate armour layers, the conical surface at the head of the groyne needs to use a special trapezoidal fence plate or other artificial blocks that can be placed randomly, such as dolosse or accropode, etc.

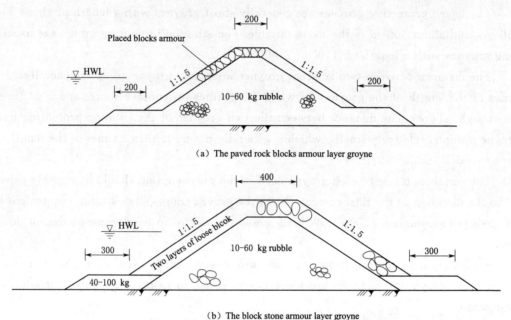

Fig. 5.3.4 The cross-section of the riprap slope groyne
(unit in elevation: m, scale: cm)

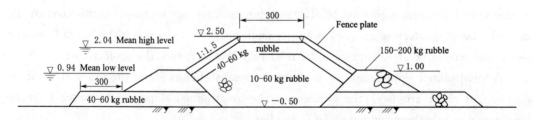

Fig. 5.3.5 The cross-section of the fence plate armour layer groyne
(unit in elevation: m, scale: cm)

The core part of the dam body at relatively low water depth can use bagged sand to replace stone to save investment.

According to the strength of the wave action, the minimum width of the top of the groyne is 1 - 2 m, and the width should also be considered as a channel during construction. The slope of the dam body is generally 1 : 1.5 - 1 : 2, and the slope of the dam head should be eased to 1 : 3 - 1 : 5.

Since the downstream side of each groyne is affected by the erosion of the beach, there should be at least 2 m riprap bottom protection layer outside the toe of the slope

5.3 Groynes

groyne. The width of the bottom layer should be less than 5 m if the scouring is relatively serious at the head of the groyne.

The head of groynes built in strong tidal estuary can form relatively deep scouring pits that affect the stability of the dam head. Therefore, there is also an example of using a caisson structure as the head of a slope groyne. The caisson is sunk to a certain depth below the seafloor so that the scour pit at the head of the groyne does not cause the collapse of the dam slope.

When the wave action is not strong, the USA has examples of wood, steel or reinforced concrete sheet piles for groynes. The sheet pile wall can be a cantilevered structure, or it can be a type in which spaced wood piles or reinforced concrete piles are connected to the sheet pile wall along the longitudinal direction of the groyne as a reinforced support. When the wave action is strong, the sheet pile can be made into a lattice structure. According to China's national conditions, if the local lacks stone materials and the construction conditions are suitable, the groyne scheme with reinforced concrete sheet pile structure as shown in Fig. 5.3.6 can be adopted.

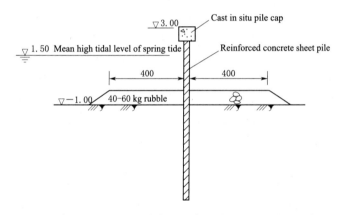

Fig. 5.3.6 The cross-section of the reinforced concrete sheet pile groyne
(unit in elevation: m, scale: cm)

In the First Phase of Yangtze Estuary Deepwater Channel Regulation Project, a new reinforced concrete semi-circular structure (Fig. 5.3.7) is used for a groin between its south and north leading jetties. The prefabricated semi-circular structure consists of a semi-circular arch ring and a base plate, which is placed on a rock-throwing base bed. This new structure has the advantages of low wave force, good stability performance and economic section, low foundation stress, being suitable for soft foundation; easy construction, resisting wave attack after semi-circular members are placed and good landscape effect.

The radius of the semi-circular member is 4.8 m, the height is 6.5 m, the thickness of the arch ring is 0.8 m, the thickness of the bottom plate is 1.3 m and the longitudinal length is 3.2 m. There are 3 rows of 66,300 mm vent holes on the outside of the arch ring

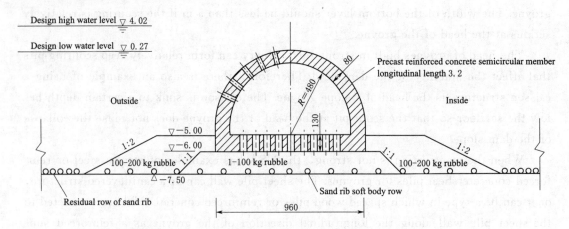

Fig. 5.3.7 The cross-section of the semi-circular structure groyne
(unit in elevation: m, scale: cm)

of each member, and 4 rows of 300 mm vent holes are provided on the bottom plate. Each member weighs 180 t and can be placed by a 200 t crane ship.

In areas with high velocity, the bottom protection structure is an important part of the groyne building. In the Yangtze Estuary Regulation Project, the bottom protection structure of the groyne mainly uses a soft sinking mattress composed of geotextile, sand, and concrete blocks. The soft sinking mattress is laid on the bottom surface to prevent scouring of the seabed by the current during the construction and service periods. From the cross-section of the groyne shown in Fig. 5.3.6, it can be seen that the sand-rib soft mattress with 1.5 m interval is used under the dam body, and the sand-rib soft mattress with 0.5 m interval is used on both sides, each side of the remaining mattress is about 25 m wide. The remaining mattress of the dam head are 40 cm×40 cm, 16 cm thick of concrete interlocking blocks soft mattress, with a width of 50 m on each side. According to the results of the water scouring test, in order to prevent the local scour pit at the head of the groin from affecting the structural stability of the head, the length of the concrete interlocking blocks soft mattress to the front side of the dam head is extended to 100 m.

The wave force which acts on the vertical groin building as well as the semi-circular structure can be calculated by referring to the relevant wave force calculation formula for the interaction between waves and buildings.

5.4 Detached breakwaters

5.4.1 Deposition features behind detached breakwaters

After the wave is diffracted by the detached breakwater, the wave height decreases in the cover area between the detached breakwaters and the shoreline, and the capacity of longshore sediment transport also decreases. The sediment input from the upstream side

5.4 Detached breakwaters

will be deposited behind the detached breakwater, making the shoreline there gradually form a prominent sand spit, as shown in Fig. 5.4.1. As the sand spit develops, its effect will be similar to that of the groin, causing siltation of the beach on its upstream side. When the length of the detached breakwater is large enough relative to its offshore distance, the sand spit will develop into a tombolo connected to the breakwater. In the case of sufficient incoming sand from upstream, roughly speaking, when the ratio of the offshore distance X_B to the breakwater length L_B of a single detached breakwater is $1-2$, a sand spit will be formed behind the breakwater extending from the shore to the sea; when $X_B/L_B < 1$, the sand spit will develop into a tombolo.

Fig. 5.4.1 gives a set of results of model tests about the development process of the sand spit behind the detached breakwater at Kyushu University, Japan. It can be seen from that in the initial stage, that is, $1-3$ h in the test, the sand spit is in the form of a double peak, while it becomes a single peak with a blunt head after 5 h. L_B in the Fig. 5.4.1 is the breakwater length; X_B is the offshore distance of the breakwater; H_0/L_0 is the deep-water wave steepness. The beach slope is 1 : 15. The research also shows that the siltation behind the breakwater is relatively large under the action of steep waves.

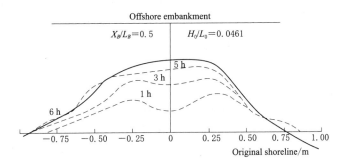

Fig. 5.4.1 The development process of the sand spit behind the detached breakwater

Similar to the case of groins, interception of incoming sand from upstream will result in insufficient sand supply to the downstream side of the detached breakwater, thus causing beach erosion.

In order to protect a certain length of shoreline, a single long dike can be built, or segmented detached breakwater can be constructed. It is usually advantageous to use a segmented detached breakwater because it can retain the sediment that enters its entrance due to lateral sediment movement along the shoreline, although it intercepts relatively less longshore sediment transport. Because the segmented detached breakwater intercepts less longshore sediment transport, the erosion of the downstream side of the beach is also weaker.

For the same total length of protected shoreline, the actual dike length of the segmented detached breakwater is shorter and less expensive to construct than that of the single long dike because the segmented detached breakwater is built intermittently with an entrance between every two short dikes. In addition, the beach utilization conditions behind the segmented detached breakwater are also more superior.

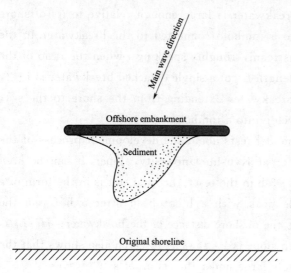

Fig. 5.4.2 The siltation behind the detached breakwater on the muddy coast

It should be noted that the siltation pattern behind the detached breakwater built on a muddy coast will be different from that of the sandy coast described above. The suspended sediment will first silt in the area with the smallest wave diffraction coefficients behind the detached breakwater, resulting in the siltation pattern as shown in Fig. 5.4.2.

5.4.2 Layout and main dimensions of detached breakwaters

The previous section roughly gives the conditions for the formation of sand spit and tombolo after a single dike has been established. In fact, the deformation of the shoreline behind the dike is related to many factors, such as water depth and wave conditions (e.g., wave height, period, and direction), beach condition (e.g., sediment diameter, shore slope), the length, top elevation and distance from the shore of the detached breakwater and the permeability of the dike.

Based on a large number of experiments and field data, researchers at the University of Delaware, USA, analyze that the relative position of the axis and the wave breaking line of the detached breakwater is a comprehensive and important factor. For a single detached breakwater, the relational expressions of the sand spit length X_S for different dike lengths L_B and offshore distances X_B are respectively given:

$$\begin{cases} X_S^* = 0.156 L_B^* & X_b^* < 0.5 \\ X_S^* = 0.317 L_B^* & 0.5 \leqslant X_b^* < 1.0 \\ X_S^* = 0.377 L_B^* & X_b^* \geqslant 1.0 \end{cases} \quad (5.4.1)$$

$$X_b^* = X_b / X_B; \quad X_S^* = X_S / X_B; \quad L_B^* = L_B / X_B$$

Where, X_b is the distance between the wave breaking line and the shoreline; X_S is the length of the sand spit protruding from the shoreline (Fig. 5.4.3).

From the Eq. (5.4.1), it can be concluded that when $X_b^* = 1.0$, that is, the detached breakwater is located at the position of the wave breaking line, $L_B = 2.65 X_B$, then a tombolo can be formed; when $X_b^* = 0.5$, that is, the offshore distance of the dike is twice of the distance between the wave breaking line and the shoreline, $L_B = 3.15 X_B$,

5.4 Detached breakwaters

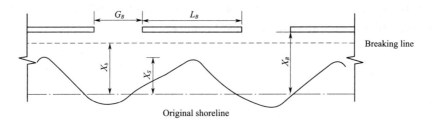

Fig. 5.4.3 The parameter diagram of detached breakwater system

then a tombolo can be formed.

For the segmented detached breakwater, the length of the sand spit X_s is also related to the width of the entrance between the two detached breakwaters G_B.

$$X_S^* = 14.8 \frac{G_B^*}{L_B^{*2}} \exp\left[-2.83\left(\frac{G_B^*}{L_B^{*2}}\right)^{1/2}\right] \quad (5.4.2)$$

$$G_B^* = G_B / X_B$$

The peak value of the Eq. (5.4.2) is at $G_B^*/L_B^{*2} = 0.5$.

Physical model tests on three segmented detached breakwaters at Cheng Kung University in Taiwan show that the parameter plays an important role in the siltation behind the dike, and the siltation effect is best when this parameter is about 0.5.

Japan is the country with the most detached breakwaters. According to the survey data of 1,552 detached breakwaters along the coast of Japan, the most common dike length is 100–110 m; 90% of the dikes are built in water depths of 5 m or less, and the most common water depth is 3–4 m; 65% of the total number of dikes have a top elevation of 1–2 m above mean sea level; 63% of the total number of dikes have an offshore distance of 20–80 m; and 60% of the total number of dikes have a tombolo behind them. In addition, it is pointed out that maintain shoreline directly opposite the entrance after the shoreline behind the dike has been deformed $G_B/L_B \leqslant 0.3$ should be taken.

Another Japanese statistical data of segmented detached breakwater project examples, linking dike length with wave length, points out that each section of dike length can be 2 to 6 times of wave length, equivalent to 60–200 m; the width of the entrance between dikes can be 20–50 m.

The TU Delft in the Netherlands presented the results of a three-dimensional dynamic bed physical model test about segmented detached breakwaters at the 25th Coastal Engineering Conference. The test showed that when the top elevation of the detached breakwater is below the tidal level, that is, a submerged breakwater, the reflux of the water formed by the wave over the top will be concentrated at the entrance between the dikes and cause scouring. Therefore, the submerged breakwater cannot play a good role in protecting the beach.

Generally, the top of the detached breakwater should be about 1.0 m high above of the 10% of the cumulative frequency of high tide or the averaged high tidal level.

5.4.3 Sedimentation amount after detached breakwaters

The prediction of sedimentation amount after each dike in a segmented detached breakwater is relevant for determining the effect of the detached breakwater system, carrying out the utilization planning of the shoreline behind the dike and estimating the degree of scouring of the downstream shoreline.

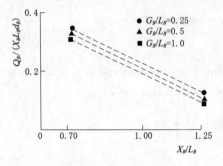

Fig. 5.4.4 The dimensionless sedimentation amount behind the dike

The relevant physical model tests in the USA yielded a graph of the relationship (Fig. 5.4.4) between the relative offshore distance X_B/L_B and the dimensionless sedimentation amount behind the dike $Q_B/(X_B L_B d_B)$ under the condition of different relative widths of entrance G_B/L_B. Q_B is the sedimentation amount behind each dike, d_B is the water depth before the detached breakwater, and other symbols are the same as in Section 5.4.2.

As it can be seen from the Fig. 5.4.4, the sedimentation amount behind the dike will decrease as the relative entrance width increases.

If the effect of relative entrance width is not considered, the following empirical Eq. (5.4.3) can be obtained by combining the data from progenitor measurements and model experiments.

$$\frac{Q_B}{X_B L_B d_B} = \exp\left(0.31481 - 1.92187 \frac{X_B}{L_B}\right) \qquad (5.4.3)$$

Eq. (5.4.3) applies to X_B/L_B from 0.5 to 2.5. When $X_B/L_B < 0.5$, the data is less, but the dimensionless sedimentation amount $Q_B/(X_B L_B d_B)$ behind the dike will not exceed 0.5.

Actual observations show that about half of the sedimentation amount occurs in the first year after the construction of the dike, and the stable siltation pattern will be completed in 4 - 5 years.

5.4.4 Overview of the headland breakwaters

Based on the study of the regular pattern of forming a dynamically balanced curved shoreline between natural headlands, we have developed the construction of headland breakwater as a new facility for coastal protection. The axis of the headland breakwater is usually arranged perpendicular to the dominant wave direction, as can be seen in Fig. 5.4.5, which is actually a form of arrangement between the detached breakwater and the groin.

Between two headland breakwaters, an artificial headland bay with a certain planar curve shape will be formed, which is compatible with the local longshore sediment transport. Since artificial headlands basically do not intercept the upstream sediment, they usu-

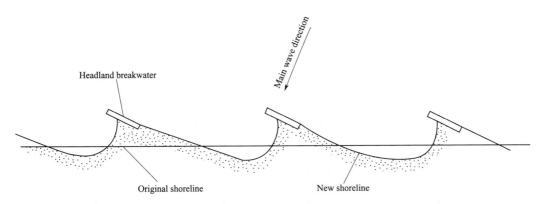

Fig. 5.4.5 The headland breakwater

ally do not aggravate the erosion of the downstream shoreline. Generally, the distance between two headland breakwaters can be slightly larger than the distance of the groins, such as the distance of headland breakwaters used in some actual projects is about 4 times of their length. Compared with detached breakwaters, headland breakwaters have the possibility of being constructed from land, which is one of their advantages.

It should be noted that in the process of forming a dynamically balanced curved shoreline between the two headland breakwaters, the beach behind the dike which grows toward the dike gradually form a tombolo. At the same time, the open section of the coast between the two dikes will be scoured so that the shoreline recedes. This phenomenon of local shoreline retreat should be fully considered in the planning and design.

In conclusion, headland breakwater is a relatively new facility to protect the coast, and its practical experience is much less than that of groin and detached breakwater. Therefore, headland breakwater is temporarily classified as a detached breakwater in this chapter, as a special form of detached breakwater. The design method of headland breakwaters needs to be developed and improved.

5.4.5 Structural types of detached breakwaters

The most basic structural type of detached breakwater is the riprap slope dike that uses large stones or various types of concrete blocks such as tetrapod, dolosse, accropode, etc. as face protection. For example, in a survey of 1,552 offshore dikes published in Japan, all the dike types are riprap slope structures.

Fig. 5.4.6 shows a typical cross-section of a detached breakwater with a large stone armour layer on the Tel Aviv waterfront in Israel. The local tidal range does not exceed 50 cm, the top of the dike is 1.75 m above the mean sea level, and the width of the dike top is 5.25 m. The slope of the sea side is 1:3, and 5-8 t boulders are used to protect the surface; the slope of the shore side is 1:1.5, and 3-5 t boulders are used to protect the surface.

Fig. 5.4.7 is a cross-section of a detached breakwater with an accropode armour layer

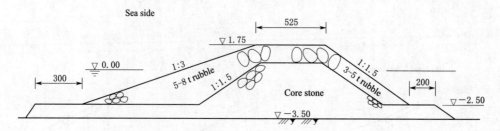

Fig. 5.4.6 The cross-section of the large stone armour layer detached breakwater
(unit in elevation: m, scale: cm)

designed for a coastal protection project in Africa. The slope of the detached breakwater is 1 : 1.5 both inside and outside, and the accropode of armour layer weighs 6 t with 400 – 800 kg stone layer under it.

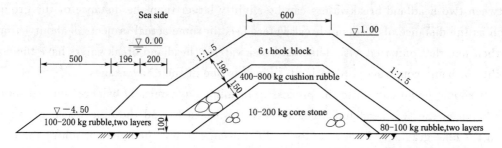

Fig. 5.4.7 The cross-section of the artificial block armor layer detached breakwater
(unit in elevation: m, scale: cm)

The new semi-circular structure is also suitable for building detached breakwaters, especially in areas which are lack of large boulders. Fig. 5.4.8 shows the scheme of a semi-circular structure detached breakwater. The design wave height of the detached breakwater is 5.1 m and the semi-circular members can be placed by a 200 t crane.

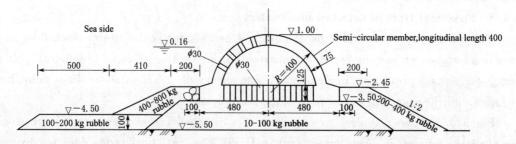

Fig. 5.4.8 The cross-section of the semi-circular structure detached breakwater
(unit in elevation: m, scale: cm)

The formula for calculating the stability weight of the slope dike armour block or artificial block and the calculation method of the wave force acting on the semi-circular dike are referred to the calculation method of the interaction between waves and buildings. The design calculation content and structural requirements of detached breakwaters are basically the same as those of breakwaters in harbor projects, and the design and construction of detached breakwaters can refer to the relevant provisions of the Ministry of Transportation's *Code of Design for Breakwaters and Revetments* (JTS 154—2018).

According to the Japanese survey, it should be noted that 13.7% of the total number of detached breakwaters had varying degrees of scattering of the armour block. Based on the results of the analysis, it is recommended to multiply the coefficient of 1.5 when the Hudson formula is used to calculate the weight of the armour block. The main reason is that the top elevation of detached breakwaters is generally lower than the top elevation of the harbor breakwater. This is basically the same as the regulation of the *Code of Design for Breakwaters and Revetments* (JTS 154—2018), which states that when the top elevation of the dike is less than 0.2 times of the design wave height above the design high tidal level, the weight of the top block should not be less than 1.5 times of the weight of the outer slope armour block.

5.5 Artificial beach nourishment

5.5.1 Overview of artificial beach nourishment technology

Artificial beach nourishment is an engineering measure which is usually used for eroded coasts, that is, sand is collected from sea or land-based sources and then filled on the beach to make up for the eroded sediment and prevent the coastline from receding. It should be said that this is the most natural and simple way of beach protection, yet this is a relatively new technical method compared to traditional revetments, groins and detached breakwaters.

Because artificial beach nourishment is generally an economical and effective solution, and the impact on adjacent coastal areas is relatively small, it has been increasingly adopted by various countries.

In 1985, the 26th International Navigation Congress had a topic group about the latest development of artificial beach nourishment technology and equipment for sandy coasts, and in 1991, the international journal *Coastal Engineering* published a special issue about artificial beach nourishment, which shows the international attention to this kind of coastal protection facilities.

The scale of beach nourishment projects can range from a few 1,000 m^3 to several 1,000,000 m^3. Large-scale beach nourishment is usually filled from the water, and the most economical method is to use the cutter suction dredger for hydraulic reclamation. Generally, the distance of the sand pits on the seabed is not more than 10 km, and it is relatively economical and reasonable if beach nourishment can be combined with the excavation of harbor basins and channels. In the late 1970s, Belgium carried out large-scale beach nourishment in the coastal area adjacent to the Netherlands, filling 8,400,000 m³ of sand on the coastline of about 8 km, and the filling sand quantity per linear meter was as high as 1,000 m³. Of course, beach conditions and natural conditions vary from place to place, so the amount of sand replenishment may also vary greatly, but roughly speaking, the average amount of sand replenishment on artificial beaches is about 100 m³/m.

When adopting a beach nourishment program, we must accept the notion that beach nourishment does not put things right once and for all. The sediment that is artificially filled on the beach will still be washed away by various marine environmental conditions, especially waves. Generally speaking, it is a sign of success that the amount of sand lost each year does not exceed 10% of the amount of sand replenishment. Practically, it is not necessary to replenish the beach every year. but usually for several years at a time, e. g., 5 – 10 years of erosion.

On a stable coast, when the beach needs to be widened or replenished with sand of suitable texture, grain size and color for the purpose of seaside recreation, the loss of sediment is certainly smaller than that of an erosional coast. However, it should be noted that a stable coast means that it is in dynamic balance. In a stable coast area, there will still be longitudinal longshore sediment transport, but for a certain shore section, the amount of sediment entering the section from the upstream of longshore sediment transport direction is equal to the amount of sediment transported to the downstream. Therefore, newly suitable filling sediment will still be gradually lost. In addition, there is also transverse sediment transport on the stable coast that can cause seasonal changes in the beach profile, thus causing the filling sediment on the beach to flow downhill.

5.5.2 Design of artificial beach nourishment

For coastal protection projects using artificial beach nourishment measures, in order to reduce the loss of filling sediment, sediment with relatively large particle size can be selected, or combined with the construction of simple marine structures. The median particle size D_{50} of the sediment required for artificial beach nourishment can be about 1.0 – 1.5 times of that on the original beach. Of course, it should be noting that sediment which is too coarse in size will not be suitable for recreational beaches. In addition to the grain size, the composition of the sediment is also an important factor in the stability of the filling sediment. Generally, the sediment obtained from the dunes on land or the surface of the sea floor is not only fine but also well sorted (poorly graded), and it is easy to be dried and lost. In contrast, the sediment obtained in alluvial rivers or shoals in the sea is often coarse and well graded, and it is relatively stable.

When longshore sediment transport is large, short and low groins or groin groups can be constructed to prevent the filling sediment from being washed away by alongshore current. Many large beach nourishment projects abroad are combined with groin systems. Detached breakwaters are particularly suitable for situations where transverse sedi-

5.5 Artificial beach nourishment

ment transport is strong and they have a significant effect on the protection of subsequent beaches. One of the concluding comments of the 26th International Navigation Congress on beach nourishment technology was that beach nourishment on sandy coasts can be combined with the construction of short and low groin groups or detached breakwaters when needed, which is an economical solution. Such a comprehensive treatment plan, if properly designed, could eventually transform the beach from erosion to accumulation. Another simpler and more economical measure is to cast submerged prisms composed of blocks or bagged sand on the low part of the beach, which can reduce the amount of sediment loss from the beach nourishment during the construction period as well as during the service period.

Beach nourishment alone, as above mentioned, has little impact on the adjacent coast because it is less disruptive to the natural balance of the coastal zone. When beach nourishment is combined with groins or detached breakwaters, it may cause erosion in the downstream section because longshore sediment transport is partially cut off, so a relatively comprehensive regulation and design must be carried out. If necessary, numerical simulation should also be conducted.

The profile design of the beach nourishment project needs to address the problem of the slope of the filling sediment. Under certain dynamic conditions, the beach slope is related to the sediment grain size. As a design profile, the slope of filling sediment above the low tidal level can be roughly parallel to the natural shore slope. In the absence of information, the slope of the filling sediment can be roughly 1 : 50, 1 : 25, 1 : 15 and 1 : 10 for D_{50} of 0.2 mm, 0.3 mm, 0.4 mm and 0.5 mm, respectively. The slope below the low tidal level can be slightly steep so that the filling sand slope can intersect with the natural shore slope. The above design profiles can be used to calculate the amount of filling sand required. However, it is impractical and unnecessary to throw the above-mentioned slope on the beach surface and underwater. The actual construction profile can be cast into a relatively wide water platform and a relatively steep slope, and then the slope will quickly become steeper and the water platform become narrower under the wave action, gradually approaching the design profile.

If sediment with a finer median particle size than that on the natural beach has to be used, it is also necessary to consider overfilling the sand because the fine-grained sediment flows into the sea is in large quantities. The required sand-filling profile can still be maintained even after large amounts of fine-grained sediment have been lost.

If the coastal section for artificial beach nourishment is too short, the widened beach will be relatively more prominent than its adjacent coasts, which may cause local wave energy concentration and increased longshore current velocity, and make this section of coast relatively prone to be scoured. Therefore, the coastal section with artificial beach nourishment generally should not be too short. According to foreign examples, the shortest is 600 – 1,000 m.

5.5.3 Calculation of overfill rate of artificial sediment replenishment

In the previous section, it has been pointed out that sediment with relatively coarse median grain size and suitable gradation should generally be selected for beach filling to reduce the loss of filling sand. However, in practice, it may be difficult to find suitable filling near the project site. When it is necessary to use filling with relatively fine median particle size and poor gradation, overfill is required to compensate for the loss of fine-grained sediment under the sorting of wave action. For this purpose, it is necessary to calculate and determine the overfill rate, which is the actual volume of filling sand required to retain 1 m³ of filling sand on the beach, and it is mainly related to the particle size distribution of both the original sand on the local beach as well as the artificial filling sand.

The calculation method in the early years assumed that due to the sorting of the wave action on the sediment, only the part of the filling with the same particle size as that in the original local sand would be retained, that is, the final particle size distribution curve of the mixed local and filling sand on the beach after artificial sediment replenishment would be the same as the particle size distribution curve of the original local sand. Based on the above assumption, the derived formula for the overfill rate was used in the USA *Shore Protection Manual*.

Obviously, the above assumption is not appropriate for relatively coarse and stable sand in the filling. When it is assumed that the relatively fine grain size of the sediment is lost under the sorting of wave action, a new and more reasonable formula for the overfill rate can be derived, and the result of the calculation according to the new formula is plotted in Fig. 5.5.1 The calculation diagram of the overfill rate R_A, which is the calculation diagram of the overfill rate.

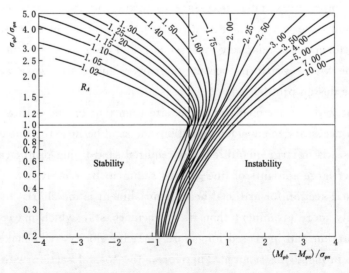

Fig. 5.5.1   The calculation diagram of the overfill rate R_A

The parameters M_φ and σ_φ in the figure are the mean and standard deviation of the sediment particle size expressed in φ, respectively. The subscripts n and b denote the local and filling sand, respectively. The relationship equation between φ and the sediment particle size D (mm) is:

$$\varphi = -\log_2 D \tag{5.5.1}$$

The corresponding relation with D is shown in Table 5.5.1.

Table 5.5.1   Correspondence between φ and D

φ	−3.0	−2.5	−2.0	−1.5	−1.0	−0.5	0.0	0.5	1.0	1.5	2.0	2.5	3.0
D/mm	8.000	5.660	4.000	2.830	2.000	1.410	1.000	0.707	0.500	0.345	0.250	0.177	0.125

M_φ and σ_φ can be calculated by the following two equations:

$$M_\varphi = (\varphi_{84} + \varphi_{16})/2 \tag{5.5.2}$$

$$\sigma_\varphi = (\varphi_{84} - \varphi_{16})/2 \tag{5.5.3}$$

Where, φ_{84} and φ_{16} are the φ values corresponding to 84% and 16% on the cumulative distribution curve of sediment particle size, respectively.

In Fig. 5.5.1, a vertical line drawn upward from a value of $(M_{\varphi b} - M_{\varphi n})/\sigma_{\varphi n}$ on the horizontal axis intersects with a horizontal line drawn to the right from a value of $\sigma_{\varphi b}/\sigma_{\varphi n}$ on the vertical axis, and the required overfill rate R_A can be obtained from the intersection point in the isopleth of R_A. The lower left side of the cluster of R_A lines in the figure indicates stable artificial beach nourishment, while the lower right side of the cluster of R_A lines indicates unstable artificial beach nourishment.

5.5.4 Sand by passing

Sand by passing can be considered as a special case of artificial beach nourishment. When artificial structures such as tidal inlet or jetty intercept or partially intercept longshore sediment transport, the downstream side of the beach will be scoured. For the eroded beach on the downstream side, artificial beach nourishment measures can be used. If the sediment accumulated in the upstream is directly transported to the downstream beach by means of mechanical transportation, such as near shore suction pump and silt discharge pipelines, which is called sand by passing.

When determining the capacity of the suction mud pump, we not only need to know the annual amount of longshore sediment transport, but also should consider the unevenness of its daily distribution.

Due to the numerous and complex marine environmental conditions associated with coastal engineering, so far, any coastal engineering may require some adjustments or partial modifications in its implementation. When artificial beach nourishment is used, it is obviously more flexible and easier to adjust compared to the construction of fixed coastal structures.

Before and after the beach nourishment project, systematic measurements of beach evolution should usually be carried out to accumulate information and continuously improve future replenishment plans.

5.6 Biological coast protection

The biological coast protection project refers to a measure of planting vegetation to resist the impact of external dynamic factors such as tides and waves in coastal protection engineering in a low-energy environment, thus the coastal embankment can be protected from erosion. It is characterized by:

(1) It often cannot be calculated by quantitative design theories and methods, but rather a lot of practice and judgment.

(2) It requires careful management.

(3) It has certain requirements for the environment and has seasonal characteristics.

(4) It can improve the environment and contribute to ecological balance.

(5) It has economic value and can be used comprehensively.

(6) It has low investment and quick results.

The bank protection function of the biological coast protection project is mainly reflected in the ability to eliminate wave energy, slow flow, and promote siltation and foundation consolidation. The friction resistance of the biological coast protection project and the reduction of water depth accelerate the dissipation of wave energy propagating to the shore. Slowing flow reduces the sediment carrying capacity of the water body and causes siltation, at this time, the water depth becomes smaller to facilitate the further dissipation of wave energy. The consolidation effect of plant roots improves the resistance of coastal regions to scouring. In this way, plant projects play the role of beach protection and siltation promotion.

The flora of the biological coast protection project to protect the embankment are water plants, grasses, shrubs, and trees. When implementing the plant coast protection project, the key is to choose a plant that is highly reproductive, vigorous, widely adaptable, highly resistant, easy to manage, economically valuable, and effective in protecting the coast. The plants used for coast protection in China mainly include reed, mangrove, spartina anglica and spartina alterniflora, among which spartina alterniflora has the best effect. Spartina alterniflora is a kind of perennial grass plant with strong salt and flooding tolerance and the stem is 1 cm thick and 1 - 2 m high, which has the functions of beach protection, siltation promotion, soil improvement, beach greening and environmental improvement. In addition, it has low requirements for soil, a wide range of temperature adaptation and low requirements for water quality. Thus, within 5 years after it was introduced to China in 1979, it had been successfully tested on the beaches of Dianbai County, Guangdong Province in the south and Yexian County, Shandong Province in the north,

5.6 Biological coast protection

and the effect is very good. At present, planting has been promoted on the beaches of Jiangsu, Fujian, Shanghai, Zhejiang and Shandong and other provinces and cities. In many places, Spartina alterniflora beach protection projects have been linked together, playing an important role in protecting the coast. In addition, we should emphasize that when planning the selection of plants for beach protection projects, we should first select plants whose growth process is consistent with the conditions of external dynamic factors acting on the coast, that is, the plants mature just when the tide is high and the waves are strong, then the coast protection effect is most ideal.

Due to the advantages of the spartina alterniflora project, coupled with its lush foliage and high stem, strong natural and reproductive ability, and easy to survival, it has been vigorously promoted and applied especially in the southern region of the Yangtze River, and has received good results.

For example, the Jiangnan seawall in Cangnan County is 15 km long, 7.5 km of which is planted with 20 - 250 m wide spartina alterniflora in front of the beach, and no spartina alterniflora is planted in front of the other 7.5 km of the seawall. In Typhoon 9414, it was observed that the overtopping water body in the dike section without spartina alterniflora was several tens of meters high and the dike break was 30 - 50 m wide, while the dike section of the beach planted with spartina alterniflora was not damaged. Table 5.6.1 shows the measured data about the wave dissipating effect of spartina alterniflora in front of the dike in the test station.

Table 5.6.1 Measured data from an automatic wave measuring instrument at the test station

Working conditions		Maximum wave height/m	
		No grass	Spartina alterniflora 250 m
Typhoon 9414 (8 August 1994)	150 m in front of the dike	1.55	0.63
	10 m behind the dike	0.8	0.03
Typhoon 9414 (21 August 1994)	150 m in front of the dike	2.04	1.16
	10 m behind the dike	1.42	0.08

From Table 5.6.1, spartina alterniflora has a good wave dissipating effect. According to the research for many years, it has a significant effect on silt promotion and soil consolidation, etc. This plant should be vigorously promoted for wave prevention measures.

Spartina anglica is also a biological beach fixing measure, which is planted more in Qidong, Sheyang and Dafeng Counties of Jiangsu Province, and can play the role of beach fixation and siltation promotion. However, the effect of planting spartina anglica on strongly eroding beaches is not yet satisfactory. For example, the spartina anglica planted on Lvsi Beach was all destroyed in Typhoon 9414. Through the investigation of spartina anglica planted in Jiangsu Province, it is considered that the maximum water depth of spartina anglica planting should not exceed 2.5 m, and it is also not suitable for planting in shellfish breeding grounds.

Exercise

Q1　Fill in the blanks: The difference between the dynamic environment of river reservoir revetment and coastal revetment lies in (　　).

Q2　Multiple choices: According to the layout arrangement and the action of regional combined waves and currents, revetment is divided into direct revetment and indirect revetment. The following are indirect revetment buildings: (　　).

　　A. Groins　　　　　　　　B. Submerged dams
　　C. Embankments　　　　　 D. Seawalls

Q3　Fill in the blanks: The positions of the sloping breakwater foot protection structure are in (　　) and (　　).

Q4　Multiple choices: The main forms of foot protection commonly used in seawall and revetment projects are (　　).

　　A. riprap foot protection　　　B. artificial block
　　C. sinking mattress　　　　　 D. bagged sand

Q5　Short answer: Introduction of calculation method about riprap foot protection weight of seawall or revetment.

Q6　Short answer: Introduction of the characteristics and conditions of use of the sinking mattress foot protection.

Q7　Fill in the blanks: In order to prevent the seabed in front of the seawall or the revetment from being scoured by the wave current, it is necessary to take measures (　　).

Q8　Fill in the blanks: Groin consists of the dam head, dam body and dam root, and its functions are (　　) and (　　).

Q9　Choices: In river regulation, in order to make more sediment deposited in the dam field area, a single groin with (　　) layout arrangement is more suitable.

　　A. the upward cantilever type　　B. the downward cantilever type
　　C. the forward cantilever type　　D. all of the above

Q10　Judgment question: For the coast which is mainly influenced by wave, it is most suitable to adopt the layout arrangement of groins perpendicular to the shoreline.

Q11　Judgment question: The groin length must be extended to the surf zone and outside the strong alongshore current zone, otherwise there is no role for shore protection.

Q12　Short answer: When using the groin group to protect the coast, how to determine the general distance between the groins?

Q13　Fill in the blanks: The elevation of the dam head for river regulation should use (　　); the elevation of the dam head for shore protection should use (　　); the

Exercise

elevation of the dam head for siltation promotion should use ( ).

Q14 Fill in the blanks: Longitudinal groins consist of ( ), ( ), and their functions are ( ), ( ).

Q15 Fill in the blanks: The layout types of longitudinal groins are ( ), ( ).

Q16 Short answer: Introduce the influencing factors of intermittent layout of longitudinal groins.

Q17 Fill in the blanks: The main factors affecting the wave dissipating effect of submerged breakwaters are: ( ), ( ), ( ), ( ).

Q18 Fill in the blanks: The purposes of plant coast protection are: ( ), ( ).

Q19 Fill in the blanks: The design of artificial beach nourishment for coastal projects mainly includes ( ), ( ), ( ), ( ).

Q20 Fill in the blanks: At present, the main ecological slope protection measures are ( ), ( ).

Q21 Fill in the blanks: The purposes of artificial beach nourishment are: ( ), ( ); artificial beach nourishment requires regular sand replenishment, and sand loss ( ) per year is successful; special attention should be paid to the ( ) sand transport and ( ) sand transport of the coast, and for the artificial beach of the ( ) coast, attention should be paid to beach muddy problems.

Q22 Fill in the blanks: The design of artificial beach nourishment for coastal projects mainly includes ( ), ( ), ( ), ( ).

Q23 Argumentative Questions: When designing an artificial beach project, it is necessary to consider the form of the shoreline and select the appropriate sediment grain size, building, filling sand slope and filling sand length, so how are these four points generally determined?

Chapter 6 Breakwater Engineering

6.1 Introduction

6.1.1 Function of breakwater

Harbors built on open shores, bays or islands are usually covered by breakwaters to create sheltered water areas. The function of the breakwater is mainly to defend the harbor from wave attack. No matter what kind of arrangement of the breakwater, the wave height in the sheltered area behind the breakwater is smaller than the incident wave height outside the entrance. The breakwater can provide smooth and safe berthing and operating waters for ships, and also improve the design conditions of wharf, revetment and other buildings in the harbor.

On the sandy coast, the breakwater connected to the shoreline can stop the longshore sediment transportation from entering the port area. On the muddy coast, with the decreasing sediment concentration in water for larger distance from the shoreline, the breakwater extending to a certain water depth can reduce the sediment entering the port area from harbor entrance by tidal effects. In addition, the breakwater could prevent the strong current or drift ice along the coast from invading the port area.

The training jetty or sand prevention dike built at the estuary or tidal inlet could also reduce the wave incidence, although it is primarily applied to guide the water flow, or to intercept drifting sand along the coast. These two types of dikes are basically the same as breakwaters in the main aspects of structural design.

The special function of breakwater makes it an important part of harbor engineering. The breakwater directly bears the strong impact of the incident waves from the open sea areas. If the breakwater is damaged in storm surge, this will not only affect the normal operation of the port, but may also cause accidents to ships berthed in the port area and damage to buildings in the port. Therefore, forthe breakwater construction project, it is necessary to carry out reasonable and thorough design and construction on the basis of fully mastering all kinds of relevant data to ensure its necessary safety.

6.1.2 Type of breakwater

For breakwaters, there are usually two classification methods: one is according to

6.1 Introduction

the planar form, and the other is according to the structural type.

When classified in planar form, if one end (embankment root) of the breakwater is connected to the shoreline, it is called the jetty. When both ends of the breakwater are not connected to the shoreline, it is called the island breakwater. The axial line of the island breakwater is generally parallel to the coastline. When the island breakwater is applied in harbor engineering, it is usually located in deep water in order to form the sufficient port area. When the island breakwater is applied for coastal protection, it is also called the offshore breakwater and located in shallow water near the coast in order to prevent coastal erosion and promote the siltation of the beach surface.

When classified in structural type, breakwaters can be divided into the following categories: mound type, vertical type, mixed type, permeable type, floating type, compressed air type and hydraulic type. The cross-sections of various structural types are shown in Fig. 6.1.1.

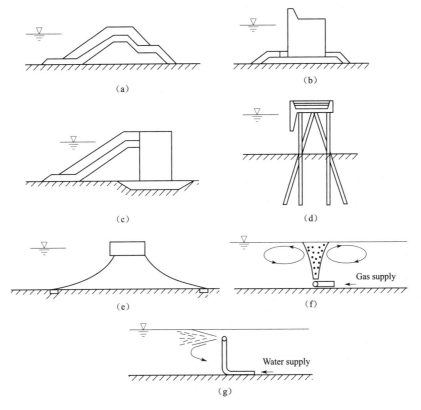

Fig. 6.1.1 The cross-sections of various structural types of breakwaters

6.1.3 Structure type and applicable conditions of breakwater

1. Mound breakwater

Mound breakwater is the oldest type, which mainly uses natural blocks, concrete blocks and various special-shaped artificial blocks as the armour.

195

Chapter 6　Breakwater Engineering

In areas where large stones can be mined, it is usually an economical scheme to use natural block stones as the protective surface, which has the advantages of solidity and durability. The stone protective surface includes riprap, placing block stones, installing block stones, dry masonry strip stone and mortar block stones.

(1) Riprap embankment with ungraded block stones. For the riprap embankment, in addition to the graded riprap embankment with large stones as the armour and smaller stones as the cushion and the core, there is also the ungraded riprap embankment with the whole embankment section built with one specification of stones. The embankment type with ungraded stones is usually used in areas with poor construction technical conditions. Since the outer side layer of the breakwater also contains light-weight rock blocks, a gentle slope is usually adopted in the strong wave action area above and below the design tidal level to maintain stability. In the upper and lower parts outside this zone, a steeper slope can be used to reduce the cross-sectional area of the embankment. Fig. 6.1.2 shows an example of a breakwater section in Delaware State, USA, which is filled with the block stones from 230 kg to 7 t.

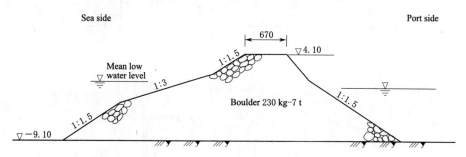

Fig. 6.1.2　Riprap embankment with ungraded rock blocks
(unit in elevation: m, scale: cm)

(2) Three-layer riprap embankment. The standard type of the three-layer riprap embankment section (that is to say, this breakwater section is composed of the armour, cushion and core) is shown in Fig. 6.1.3. Two layers of block stones with weight W are used for the armour layer. This section is applicable to the case where the water depth d is greater than 1.3 times the significant wave height H_s, that is, the design wave height is not broken. If $d < 1.3 H_s$, under the action of broken waves, the outer armour layer should extend downward until it intersects with the horizontal block stone layer above the seabed.

For the "three-layer" riprap embankment, the outer (seaward) slope can also be made into a composite slope as shown in Fig. 6.1.3. The previous experiment shows that this composite slope has better stability of the block stone.

(3) Wide-berm riprap embankment. The wide-berm riprap embankment built by the United States in 1988 at St. George's Port in the Bering Sea is shown in Fig. 6.1.4. The

6.1 Introduction

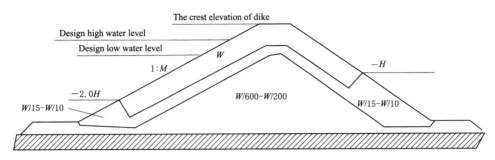

Fig. 6.1.3 Three-layer riprap embankment

local significant wave with 50-year return period is 10.4 m, and the peak period of the spectrum is 18 s. After extensive hydraulic model tests, it is determined to adopt this section type. The berm width is 16.8 m, and the weight of block stones in outer layer is 2 to 10 t. The advantage of the wide berm section is that the wave energy could be dissipated through the high void ratio berm, and the stable weight of the outer layer riprap is much smaller than that of the common "three-layer" riprap embankment. In the case of the same elevation of the embankment top, its wave overtopping discharge is also small, but the riprap volume of its section is larger than that of the "three-layer" riprap embankment. The characteristic of this type of embankment is that it allows the outer section of the dike to be deformed by the wave action until the outer slope forms a dynamic equilibrium profile.

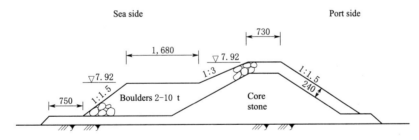

Fig. 6.1.4 Wide-berm riprap embankment
(unit in elevation: m, scale: cm)

(4) Mound breakwater with placing block stone armour. The revetment of slope embankments in China seldom adopts the form of two layers of rock blocks. Generally, one layer of stone blocks is placed, and the long axis direction of stone blocks is perpendicular to the slope surface. The weight required for placing stone blocks is lighter than that for riprap. Fig. 6.1.5 is a cross-sectional view of Qingdao Zhonggang West Breakwater in China. The top elevation of the original riprap of the embankment is about ±0.0 – 1.0 m, and it was built in 1966. The design wave height is 2 m. For the placing block stone armour, the weight of the block stones on the slope is 300 – 400 kg, and the weight of the block stones on the top of the embankment is more than 400 kg.

Chapter 6 Breakwater Engineering

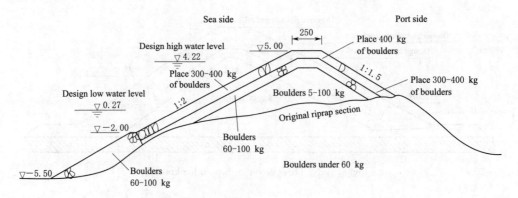

Fig. 6.1.5 Mound breakwater with placing block stone armour
(unit in elevation: m, scale: cm)

(5) Mound breakwater with dry masonry block stone armour. The dry masonry block stone armour is only applicable to the parts above the construction tidal level. And proper selection of the blocks and correction of the stone face are required. In addition to requiring the long edge of the blocks to be perpendicular to the slope of the embankment, it is also necessary to make the blocks closely embedded and stagger with each other on the slope to avoid excessive gaps. As the masonry is manually operated, the weight of the block stone should not be too small or too large. Generally, the weight of the block stone shall not be less than 50 kg, and the thickness of the masonry layer shall not be less than 30 cm.

(6) Mound breakwater with slurry block stone armour. In addition to the requirement of construction above the tidal level, the slurry block stone armour just after the construction completion shall not be scoured by the tide, so the bottom elevation of the slurry block stone armour shall be higher than that of the dry masonry block stone armour. The specification, shape and joint requirements of slurry block stone armour are lower than those of dry block stone armour. The stability of dry masonry block stone armour and slurry block stone armour under wave action mainly depends on the thickness of the armour layer rather than the weight of a single block stone. The slurry block stone armour layer must also be provided with drainage holes to remove the dynamic water pressure from the inside of the masonry slope.

(7) Riprap concrete block breakwater. In the absence of local block stone as the embankment core, embankment core can also be filled with manually prefabricated concrete blocks, and the surface layer can be protected with concrete protection blocks. For example, the Friendship Port (Port of Nouakchott) in Mauritania applied the concrete tetrahedron to build the embankment core and twisted I-shaped blocks to protect the surface because there is no local source of block stone. Fig. 6.1.6 shows the cross section of the mound breakwater formed by dumping and filling the block stone. The design wave height of the embankment is 4.6 m, but the wave transmission coefficient of the embank-

6.1 Introduction

ment body is generally large. Fig. 6.1.7 shows that the North Breakwater of the new waterway of Rotterdam Port, Netherlands in the 1970s is protected by concrete throwing rectangular blocks, with a design wave height of $H_s = 6.5$ m. The concrete blocks should not be laid regularly on the slope, otherwise, the wave run-up on the armour will be large, and the negative wave pressure under the armour may also be induced.

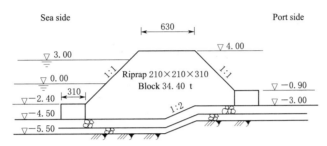

Fig. 6.1.6 Mound breakwater with throwing and filling blocks
(unit in elevation: m, scale: cm)

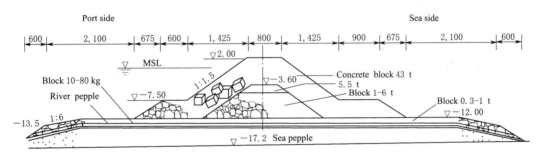

Fig. 6.1.7 Mound breakwater with riprap armour
(unit in elevation: m, scale: cm)

(8) Various special-shaped blocks for mound breakwater's armour. Since the 1950s, in order to improve the wave resistance of the slope embankment armour, reduce the concrete consumption and wave run-up on the embankment slope, a variety of types of concrete special-shaped blocks have been studied, and the starting point is to improve the void ratio of the armour layer and the locking force between the blocks. There are many kinds of blocks, but the commonly used ones in China are tetrapod, twisted I-shaped block, accropode block, Four-legged hollow block and reinforced concrete fence plate (Fig. 6.1.8). The typical cross-sections of embankment are shown in Fig. 6.1.9 to Fig. 6.1.11.

2. Vertical breakwater

The conventional vertical breakwater resists the action of wave force by its own weight, so it is generally called gravity vertical breakwater. The breakwater section is mainly composed of wall body, super structure and foundation bed. The wall body is the main part of the vertical breakwater. For the wall body, the concrete rectangular blocks or

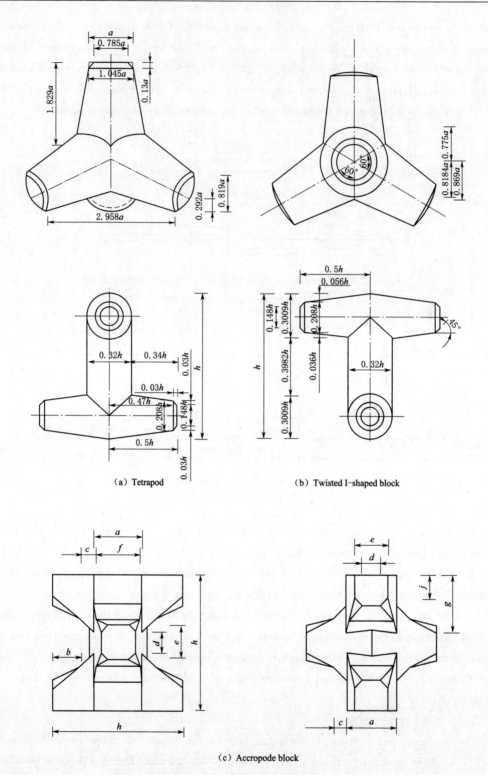

Fig. 6.1.8 (1)　Common artificial blocks

6.1 Introduction

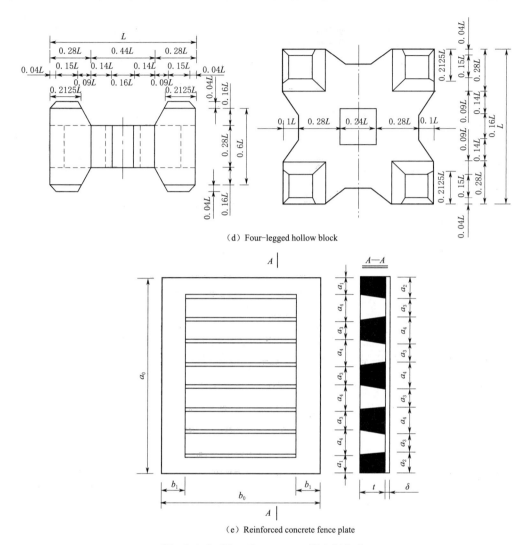

(d) Four-legged hollow block

(e) Reinforced concrete fence plate

Fig. 6.1.8 (2)　Common artificial blocks

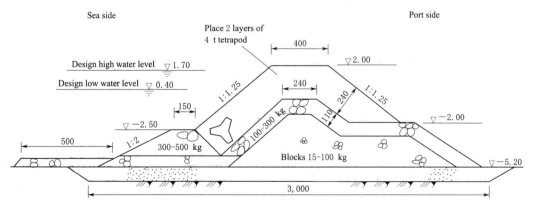

Fig. 6.1.9　Mound breakwater with tetrapod armour
(unit in elevation: m, scale: cm)

Chapter 6 Breakwater Engineering

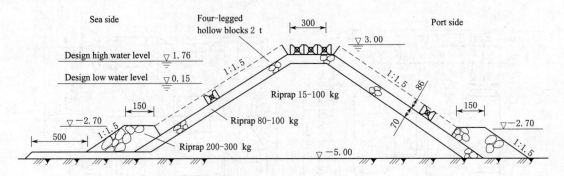

Fig. 6.1.10 Mound breakwater with Four-legged hollow block armour
(unit in elevation: m, scale: cm)

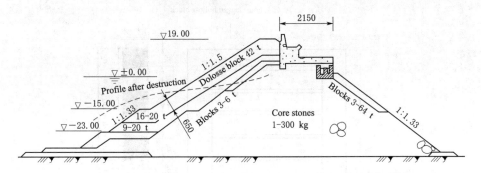

Fig. 6.1.11 Mound breakwater with twisted I-shaped block armour
(unit in elevation: m, scale: cm)

various types of hollow blocks (such as I-shaped, square-shaped and rectangle-shaped with a middle line) and reinforced concrete caissons or bottomless shafts are used generally. The concrete block wall body has the advantages of solidity and durability, but the amount of concrete is large, and the integrity is poor. It is sensitive to uneven settlement of the foundation. The construction of the block structure requires crane ship and the large workload of divers. The reinforced concrete caisson has good integrity. The caisson is filled with block stones or sand, which can save concrete consumption and speed up the construction. It is also suitable for offshore operations. However, prefabrication yard and launching facilities (such as slideway, dock or large crane equipment) are required for caisson prefabrication.

The superstructure of vertical breakwater is generally cast-in-place or prefabricated assembly concrete structure. Its seaward shape can be upright surface, arc surface or chamfered bevels. Compared with the upright surface, the arc surface can reduce the wave overtopping. The chamfered surface type has less reflection of waves, and the vertical component of wave force on the bevel is beneficial to the stability of the dike body. Therefore, the dike body's cross-sectional width of the chamfered bevel type is smaller than that of the upright surface type, but its wave overtopping volume is larger than that

6.1　Introduction

of the upright surface type.

Fig. 6.1.12 shows the cross-sectional view of general gravity vertical breakwaters. Fig. 6.1.13 shows the cross-sectional view of several vertical breakwater structures composed of special blocks.

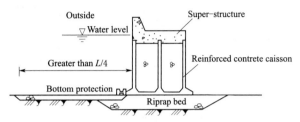

(a) Caisson type vertical breakwater

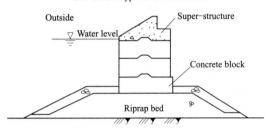

(b) Square block vertical breakwater

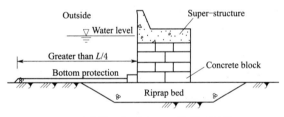

(c) Chamfered square vertical breakwater

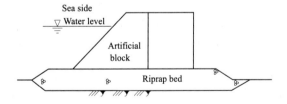

(d) Horizontal composite vertical breakwater

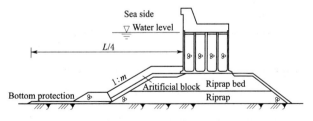

(e) Deep water vertical dike breakwater

Fig. 6.1.12　General gravity vertical breakwaters

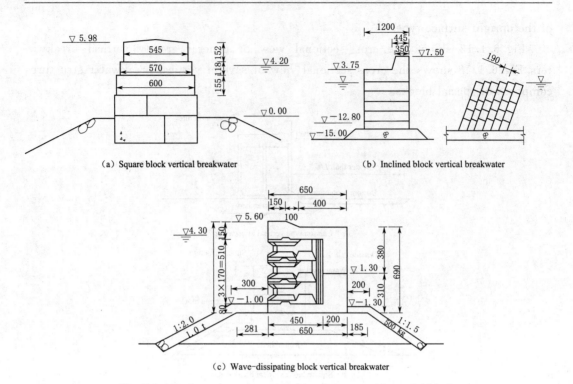

(a) Square block vertical breakwater (b) Inclined block vertical breakwater

(c) Wave-dissipating block vertical breakwater

Fig. 6.1.13 Several gravity vertical breakwaters with special blocks
(unit in elevation: m, scale: cm)

The large-diameter reinforced concrete cylinder is also used as the dike body structure of the vertical breakwater. Generally, it is a bottomless cylinder placed on the foundation bed and filled with sand and stone inside. Its working principle is basically the same as that of the general gravity type. In a few cases, part of the cylinder settles into the foundation. As for the cylinder deeply settled into the foundation, its working state is different from that of the gravity type structure because of the larger impact of the embedment of the foundation soil. Fig. 6.1.14 shows the sections of large diameter cylindrical vertical breakwater.

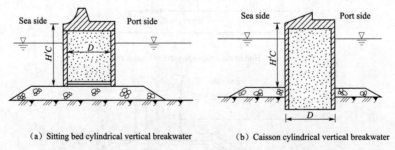

(a) Sitting bed cylindrical vertical breakwater (b) Caisson cylindrical vertical breakwater

Fig. 6.1.14 Large diameter cylindrical vertical breakwater

The foundation bed of vertical breakwater is typically a riprap foundation, which can be divided into three types: rubble mound foundation, subgrade bed or composite base bed. The selection conditions are determined according to factors such as water depth, ge-

6.1 Introduction

ology and wave elements.

3. Mixed breakwater

The embankment type with slope cover in front of the vertical wall is called the mixed breakwater, as shown in Fig. 6.1.15 and Fig. 6.1.16. Generally, it is a reinforcement measure adopted when the original design vertical embankment is damaged after long period or the vertical embankment section cannot resist the external force of incident waves. The East Breakwater of Yantai Gang was originally an inclined block vertical breakwater. Due to its long-time disrepair, the block concrete of the embankment body was seriously corroded. In order to enhance the safety of the embankment body, the outside part of the original vertical embankment was filled with stones and protected with twisted I-shaped blocks (Fig. 6.1.15). In some cases, when the design wave element is large, in order to reduce the wave force on the vertical wall, it can also be directly designed as a mixed breakwater, such as the West Breakwater of the Port of Rumoi in Japan, where the maximum wave height of 50-year return period is 13 m, so the mixed breakwater structure is adopted (Fig. 6.1.16).

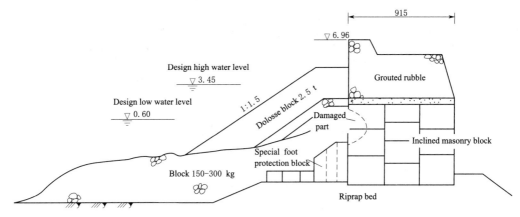

Fig. 6.1.15 The cross section of East Breakwater of Yantai Gang
(unit in elevation: m, scale: cm)

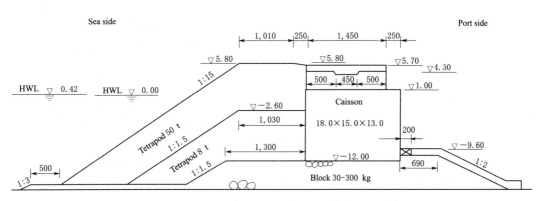

Fig. 6.1.16 The cross section of West Breakwater of the Port of Rumoi in Japan
(unit in elevation: m, scale: cm)

Chapter 6 Breakwater Engineering

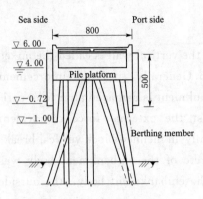

Fig. 6.1.17 Permeable breakwater
(unit in elevation: m, scale: cm)

4. Special structure breakwater

In some specific conditions, when the conventional types of the breakwater are not applicable, such as the lack of local sand and rock materials, excessive water depth and temporary/semi-permanent wave blocking measures, through the technical and economic comparison, special structure breakwater can be adopted, such as permeable type (Fig. 6.1.17), floating type (Fig. 6.1.18), high pile bearing type (Fig. 6.1.19) and lattice steel sheet pile type (Fig. 6.1.20).

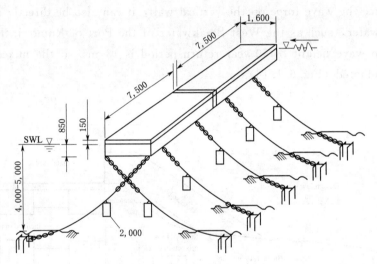

Fig. 6.1.18 Floating breakwater (unit: cm)

5. Structure selection of breakwater

In the selection of breakwater structure, various factors should be comprehensively considered. It is economical and practical to use mound breakwater when the local rock is abundant and the water depth at the breakwater location is small (generally within −7 m). When the wave height is relatively small, large stones (stone source is available), four-legged hollow blocks or reinforced concrete fence plate can be used for the armour. In the case of larger wave height conditions, artificial concrete blocks should

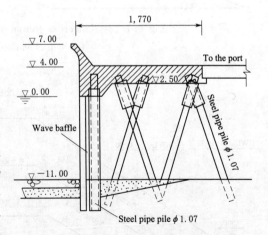

Fig. 6.1.19 High pile bearing cap type
vertical breakwater
(unit in elevation: m, scale: cm)

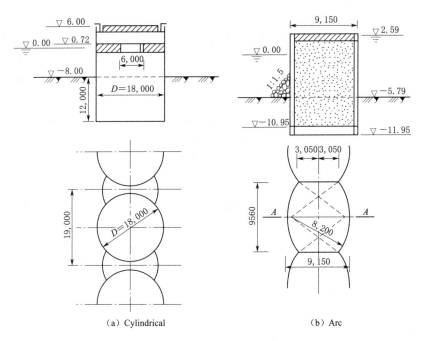

(a) Cylindrical   (b) Arc

Fig. 6.1.20 Lattice steel sheet pile (unit in elevation: m, scale: cm)

be used for the armour. When the water depth is relatively large and the geologic condition is good, the vertical structure should be adopted. The selection of the embankment structure depends on the construction conditions (such as crane, transportation and precast yard capacity). The selection of the super-structure is related to the application requirements and the criteria of permitted wave overtopping. In the case of very large water depth, the possibility of mixed breakwater must be taken into consideration, such as increasing the height of convex foundation bed at the lower part of the embankment to reduce the concrete volume of the embankment body and the difficulty of construction and prefabrication. The special type of breakwater is only applicable to the specific hydrodynamic and application conditions. For example, the high pile bearing cap can reduce the effect of the current, and its inner side could be applicable for ship berthing. But it is not applicable under the situations where with large effect of sediment on the wharf. Floating breakwater is applicable to temporary or semi-permanent breakwater requirements.

6.2 Mound breakwaters

6.2.1 Cross-sectional scale of mound breakwaters

In *Code of Design of Breakwaters and Revetments* (JTS 154—2018) the following six types of section and their determination of the relevant scales are mainly introduced.

1. Principles of section type selection

The main section type of the mound breakwater can be selected according to the following principles.

(1) When the surface is protected by riprap, placing blocks stones or artificial concrete blocks, the cross-section shown in Fig. 6.2.1 (a) can be adopted, and underwater riprap prism should be set at the toe of slope. In the case of randomly placed artificial blocks, the riprap prism is not necessary.

(2) When the dry masonry block stone, dry masonry strip stone or slurry block stone are applied for the armour protection above the water surface, the cross-section can be adopted as shown in Fig. 6.2.1 (b). The berm should be set near the construction tidal level, and large block stones or concrete blocks can be placed on the berm.

(3) When the large wave height condition or lack of the local rocks is encountered during the construction period, and if there is enough crane capacity, the cross-section of the artificial concrete block can be used, as shown in Fig. 6.2.1 (c).

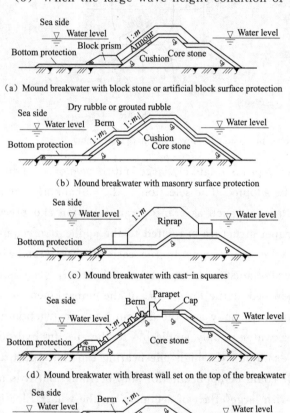

(4) When the top of the breakwater is used for access and laying pipelines, or the inner side of the breakwater is also used as a wharf, it is advisable to set up the breast wall at the top of the breakwater, as shown in Fig. 6.2.1 (d). The breast wall can be L-shaped, inverse L-shaped and arc-shaped.

(5) When stone source is abundant, block stones are used for slope protection, and the land propulsion method is used for construction, the wide berm riprap mound breakwater can be applied, as shown in Fig. 6.2.1 (e).

(6) When the sea side of the deep-water mound breakwater is protected by accropode blocks, if the slope surface is long or the number of protection blocks exceeds 18, the berms shall be set at appropriate positions on the outer slope, as

Fig. 6.2.1 Main section types of mound breakwaters

6.2　Mound breakwaters

shown in Fig. 6.2.1 (f).

2. Crest elevation of mound breakwater

The crest elevation of a mound breakwater is mainly related to the stability requirement of the harbor water areas it covers. In some old ports in China, the crest elevation of the mound breakwater is relatively low, which is less than 1.0 m above the design high tidal level. Under the effects of high tide and large waves, the wave overtopping of these breakwaters is more intense. If the crest elevation of mound breakwater is determined according to the fact that no wave overtopping occurs under the action of the highest tide level and the largest wave, it will be quite high (up to 1.0 – 1.5 times the wave height above the tidal level), which is neither economical nor necessary.

According to the investigation of ports in China and the statistics of the crest elevation values of some newly-built breakwaters in recent years, it is stipulated in *Code of Design of Breakwaters and Revetments* (JTS 154—2018) that the crest elevation of mound breakwaters shall be determined according to the use requirements and the general layout, and shall comply with the following provisions.

(1) For mound breakwaters that allow wave overtopping and have no breast wall at the top, the crest elevation of the breakwater should be set at a position not less than 0.6 times the design wave height above the design high tidal level; For the mound breakwater with block stone, four-legged hollow block or fence plate armour, the crest elevation should be set at a position not less than 0.7 times the design wave height above the design high tidal level.

(2) For mound breakwaters and riprap mound breakwaters with wide-berm platform that basically do not overtop the waves, the crest elevation of the breakwater should be set at a position not less than 1.0 times the design wave height above the design high tidal level.

(3) For the mound breakwater that basically does not overtop the waves and has a breast wall on the top of the breakwater, the crest elevation of the breast wall should be set at a position more than 1.0 times the design wave height above the design high tidal level.

(4) For mound breakwaters with high protection requirements, the crest elevation shall be determined according to the calculation of wave run-up, and the overtopping shall be controlled. The allowable overtopping shall be determined according to the protection objects and protection facilities. The specifications list the allowable overtopping provisions for sloping revetments with different protection requirements.

According to the relevant model test data, when the crest of the breakwater is $(0.6 - 0.7)H_s$ above the tidal level, the wave height behind the breakwater after overtopping is about $(0.1 - 0.2)H$.

3. Crest width of mound breakwater

The crest width of the mound breakwater shall not only meet the use requirements such as the passage of the top of the breakwater, but also ensure that it will not be damaged by waves. *Code of Design of Breakwaters and Revetments* (JTS 154—2018) in China stipulates that the crest width of the mound breakwater should not be less than 1.1 times the design wave height, and the structure should be able to place at least two rows in parallel or three artificial blocks randomly; When there is a requirement for use, it shall be determined according to the requirement.

In order to maintain stability, the required crest width and crest elevation are also related. The value of the crest width of mound breakwater mentioned above is obtained from the statistics of the mound breakwater whose crest elevations are between the value that is slightly higher than the design high tidal level and the value that is $(0.6-0.7)H_s$ above the design high tidal level.

It should be noted that the above recommended values for the crest width of the breakwater are only applicable to the general section type of the mound breakwater as shown in Fig. 6.2.1. If the top part of the embankment is not properly constructed, it may not be able to maintain stability even if the width meets the requirements. As shown in the cross-section of Fig. 6.2.1 (a), when the armour layer on the top of breakwater does not extend to the inner slope, the artificial block on the top of breakwater is easy to slide along the top surface of the horizontal cushion. When the blocks are settled in the mound breakwater, it is also subjected to large force due to impermeability to waves. For example, the Dalian fishing port breakwater is still damaged although the total width including the blocks at the top of the breakwater and the tetrapod armour in front of them is 1.4 times of the design wave height.

When land-based construction machines are used to cast the mound breakwater, the requirements of the crest width for the passing of these machines on the top of the breakwater shall also be considered. Therefore, the specification also adds in terms of the crest of the breakwater that for the mound breakwater constructed by land propulsion method, and the requirements of construction machinery on the crest width shall be considered.

4. Underwater riprap prisms

Generally, the armour block within the range of about 1 time the design wave height above and below the design tidal level is the most strongly affected by the wave. Therefore, for the cross-section with underwater riprap prism at seaward side of the embankment [Fig. 6.2.2 (a)], the top surface of the prism should be located at about 1 time the design wave height below the design low tidal level, so that the weight of the prism block applied can be relatively small. According to the model test data and the calculation results of the theoretical formula: when the top elevation of the underwater riprap prism is about 1 time the design wave height below the design low tidal level, the weight of the prism can be 1/10 – 1/5 of the weight of the rubble determined according to the formula for calculating the weight of the armour block.

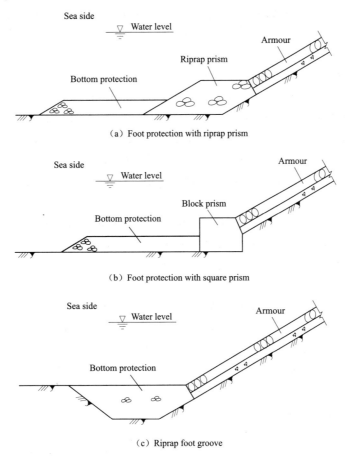

Fig. 6.2.2 Types of underwater riprap prism for foot protection

The width and thickness of the top surface of the riprap prism can be determined according to the water depth in front of the embankment and the cross-section dimension, and its width should not be less than 2 m and thickness should not be less than 1 m. For the deep-water breakwater, the width and thickness of the top surface of the riprap prism shall not be less than 5 m and 3 m respectively. According to the statistics of some projects, when $H_s \leqslant 3.5$ m, the crest width of the prism is 1.5 m; When $H_s = 4.0$ m, the crest width is 2.0 m; When $H_s = 5.0$ m, the crest width is 3.0 m. These values can be used as reference for engineering design. When artificial blocks are used as prism of the foot of slope, at least two rows of blocks of the foot of slope shall be kept, and the outer side of the block should be protected by stone blocks as top support, and the height can be half of the thickness of the block.

The role of underwater riprap prism is not only to reduce the number of the blocks of main armour, but also to support the armour to prevent its sliding and protect the bottom protection blocks from scouring. If the thickness of the prism is thin or there is the lack of large stones, the riprap of the prism can also be replaced by concrete or mortar blocks [Fig. 6.2.2 (a)]. If the armour needs to be supported and the water depth is insufficient at low tidal level, the lower end of the armour can also be extended to the sea side to form a prism [Fig. 6.2.2 (b)].

5. Armour blocks

If the water depth is large, the weight of the armour block between $(1.0 - 1.5)H$ below the design low water level is $1/10 - 1/5$ of the value determined by the formula for calculating the weight of the armour block. The weight of the armour block below the design low water level greater than $1.5H$ can be $1/15 - 1/10$ of the value determined according to the formula for calculating the weight of the armour block.

6. Berm

The platform on the mound breakwater and revetment slope is also called the berm. The berm can be used as a construction channel, reduce the wave run-up and increase the overall stability. According to the specifications, the elevation and width of the berm shall meet the following requirements.

(1) For the mound breakwater with berm due to construction needs, the elevation of the berm shall be determined according to the construction conditions, and the width of the berm shall not be less than 2 m.

(2) For the deep-water mound breakwater with berm, the elevation of the berm should be set at a position no less than 0.5 times the design wave height below the design low tidal level, and the width of the berm should at least meet the requirements of random or regular placement of three artificial blocks.

(3) For the mound breakwater with berm to reduce the wave run-up, the berm elevation should be set within 0.5 times the design wave height above and below the design high tidal level, and the width should be 0.5 - 2.0 times the design wave height.

7. Width of the mound breakwater

For the cross section of the cast-in square [Fig. 6.2.2 (c)], the wave penetration is large, so the width of the breakwater should not be too narrow to avoid affecting the stability of the water surface in the harbor. The width at the design high tidal level is usually required to be no less than 3 times the design wave height.

8. Top elevation and shoulder width of slope

For the section with wave barrier breast wall on the top of the breakwater, the elevation of the top of the wave barrier breast wall is generally set at the position not less than 1 time of the design wave height above the design high tidal level. When the armour surface in front of the breast wall is block stone, single-layer four-legged hollow block or fence plate [Fig. 6.2.1 (d)], the top elevation of the slope is generally still set at 0.6 - 0.7 times the design wave height above the design high tidal level, and the width of the

6.2 Mound breakwaters

slope shoulder in front of the wall should not be less than 1.0 m, and at least one row of surface protection blocks can be placed in the structure.

The sloping armour surface in front of the wave barrier breast wall is randomly or regularly placed with tetrapod, twisted I-shaped block, accropode block and other artificial block types, which are generally applicable to the case of large wave height. According to the analysis results of the engineering model test data, when the height and width of the artificial block slope in front of the wave retaining wall are low and narrow, the wave pressure on the wave barrier breast wall has the trend to increase due to the intensive wave breaking on the slope, and the wave barrier breast wall is not easy to be stable. Therefore, it is generally required that the elevation of the top of the slope should not be lower than the elevation of the wave barrier breast wall, and at least two rows of interconnected artificial blocks should be placed within the range of the slope shoulder in front of the wall, as shown in Fig. 6.2.3 (a) and Fig. 6.2.3 (b).

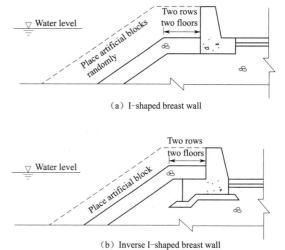

Fig. 6.2.3 Artificial block of wave wall

The calculation formulas of wave force on the wave barrier wall at the top of the mound breakwater have been given above, and these formulas can be used for the case where the sloping armour surface in front of the wave barrier breast wall is block stone or single-layer four-legged hollow blocks [Fig. 6.2.1 (d)]. When the mound breakwater has the type as shown in Fig. 6.2.3, the wave force (lateral pressure and buoyancy force) acting on the wave barrier wall can be multiplied by a reduction factor of 0.6.

Making an arc-shaped or trapezoidal wave deflecting mouth (Fig. 6.2.4) on the top of the wave barrier breast wall can effectively reduce the overtopping amount, but also increase the wave force which is unfavorable to the stability of the wave barrier breast wall. For the mound breakwater with arc-shaped breast wall [Fig. 6.2.4 (a)], the arc surface of breast wall can smoothly connect with the outer slope surface. Its length on the slope surface should not be less than 1 m and its thickness should not be less than the slope protection surface layer to ensure the stability of arc-shaped breast wall. For the breast wall of trapezoidal wave deflecting mouth, in order to increase the stability against sliding of the breast wall, a chiseling wall can be set under the bottom plate of the breast wall [Fig. 6.2.4 (b)]. When the depth of the chiseling wall is 0.5 - 0.8 m, the friction coefficient f between the wave barrier wall and riprap can be increased from 0.6 to

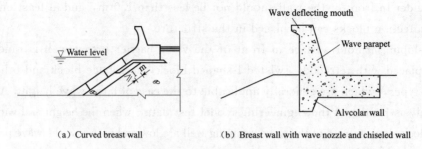

(a) Curved breast wall　　　　(b) Breast wall with wave nozzle and chiseled wall

Fig. 6.2.4　Breast wall with wave nozzle

about 0.8.

The berm top elevation of the wide-berm mound breakwater can be set at 1.0 - 3.0 m above the design high tidal level and the width of the berm should be 2.3 - 2.9 times the design wave height and should not be less than 6.0 m.

9. Slope gradient

The slope gradient of mound breakwater with various types of armour can generally be selected within the following ranges: 1 : 1.5 - 1 : 3.0 for cast-in or placed blocks; 1 : 1.5 - 1 : 2.0 for dry masonry or slurry blocks; 1 : 1.25 - 1 : 1.5 for placed artificial blocks; 1 : 1 - 1 : 1.25 for riprap-fill squares; 1 : 0.8 - 1 : 2.0 for dry masonry blocks; for the wide-berm riprap mound breakwater, the slope gradient above and below the berm can be 1 : 1.5 - 1 : 3.0 and 1 : 1.0 - 1 : 1.5 respectively; the slope of the surface should not be milder than 1 : 1.5 for artificial block armour. Of course, for the needs of the overall stability of the foundation, the slope of the artificial block armour surface for the mound breakwater should not be slower than 1 : 1.5, and the relatively mild slope such as 1 : 2 can be used.

6.2.2　Construction of mound breakwaters

For the structure of the mound breakwaters, the embankment core materials of the mound breakwaters, the placement requirements of the armour blocks, and the constructional requirements of the bottom protection are introduced.

1. Filling material for the core of the mound breakwater

The core of the breakwater shall be selected according to the local conditions. Generally, the core of mound breakwater shall be composed by 10 - 100 kg block stone, or the mixture of hill-excavating stone can be used. The weight of hill-excavating stone filler can be less than 300 kg, and for deep-water mound breakwater, it can be less than 800 kg. The hill-excavating stone shall be properly graded, and the content of 1 - 10 kg block stone and particles below 1 kg shall be less than 10%. For the areas lacking of local stone sources, substitute materials such as bagged sand can be used, and the corresponding design of the embankment core structure shall meet the technical requirements for the application of geosynthetics in port and waterway engineering.

6.2 Mound breakwaters

2. Bottom protection

When the mound breakwater is constructed on foundations that can be scoured by wave action, the bottom protection layer should be set in front of the toe of the slope, as shown in Fig. 6.2.5. According to the flow velocities at the bottom and along the breakwater which are induced by the waves in front of the breakwater, the bottom protection structure can adopt riprap bottom protection or geotextile soft mattress. The width of the bottom protection block stone layer (calculated from the toe of the slope outwards) is generally 5 – 10 m for the embankment body section and 10 – 15 m for the embankment head section. Two layers of bottom protection block stone can be used, and the thickness should not be less than 0.5 m. For sandy seabed, a gravel layer or geotextile filter layer with a thickness of no less than 0.3 m should be set under the bottom protection block stone layer. The stable weight of the bottom protection block stone can be determined according to the maximum wave-induced bottom velocity U_{max} in front of the breakwater, and U_{max} can be calculated according to Eq. (6.2.1).

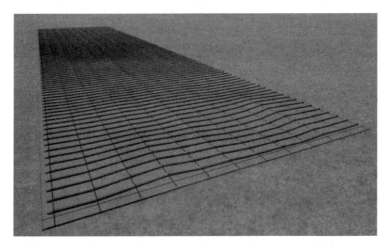

Fig. 6.2.5 Sand rib soft mattress

$$U_{max} = \frac{\pi H}{\sqrt{\frac{\pi L}{g} \sinh \frac{4\pi d}{L}}} \qquad (6.2.1)$$

When U_{max} is 2 m/s, 3 m/s, 4 m/s, or 5 m/s, the stable weight of the bottom protection block is 60 kg, 150 kg, 400 kg, or 800 kg respectively. If the scouring flow velocity in front of the breakwater is less than 2.0 m/s, the geotextile sand rib soft mattress can be used; if it is more than 2.0 m/s, the interlocking block soft mattress can also be used, as shown in Fig. 6.2.6.

For foundations that can be scoured by wave action, neither the armour blocks nor the large blocks of submerged prisms should be thrown directly on the seabed. Instead, a bedding layer of 10 – 100 kg should be laid on the seabed first. And the thickness of this

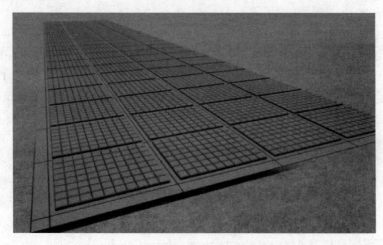

Fig. 6.2.6 Interlocking block soft mattress

bedding layer should not be less than the thickness of the bottom protection layer (Fig. 6.2.1).

3. Structural requirements for armour blocks

In the cross-section design of mound breakwaters, it is necessary to determine the weight of block or stone for each part of the embankment. The determination methods of the weight of the armour block on the outer slope and the underwater riprap prism block have been given above. The principles for determining the weight of other blocks or stones will be discussed below. When the weight of twisted I-shaped block is more than 20 t, or the weight of accropode block and tetrapod is more than 40 t, appropriate strengthening measures shall be taken.

For the weight of the armour block on the inner slope, when the wave is allowed to overtop, the same weight as that of the armour block on the outer slope shall be used for the block from the top of the breakwater to the design low tidal level. Under the design low tidal level, generally, the same weight of block stone as the outer armour cushion can be used, but no less than 150 – 200 kg, and it should be double-checked according to the wave conditions on the inner side of the breakwater. If the wave is not allowed to overtop, the weight of the inner slope armour block stone shall be determined according to the wave conditions of the inner side of the breakwater, and the block stone with the same weight as the outer slope armour cushion can be used generally.

The block weight of the top of the breakwater is generally the same as that of the outer slope. Only when the crest elevation of the breakwater is low (less than 0.2 times the design wave height above the design high tidal level), its weight should be increased to about 1.5 times of block weight of the outer slope. This is because the abroad model test of the mound breakwater with wave overtopping shows that when the top of the breakwater is about $(0.05 - 0.2)H$ above the tidal level, the stability of the rock block on the top of the breakwater is the worst.

The weight of the block stone of the armour cushion (Fig. 6.2.1) shall not be less than 1/40 – 1/20 of the weight of the armour block, and the stability under the wave action during the construction period shall be double-checked. The thickness of cushion block stone is generally taken as the thickness of two layers of block stone, which is calculated according to Eq. (6.2.1).

4. Material requirements

With regard to the stone for the construction of breakwater, the following requirements should normally be met.

(1) Strength after immersion in water. For the armour block and the subgrade bed block to be tamped, it shall not be less than 5,000 kN/m^2, and for the cushion block and the subgrade bed block not to be tamped, it shall not be less than 3,000 kN/m^2.

(2) Not flaky, not heavily weathered or cracked. However, for the core stone, the requirements can be appropriately reduced according to the specific situation.

For the concrete and reinforced concrete components of the breakwater, the frost resistant grade shall be selected according to the provisions in the relevant *Code for Design of Concrete Structres* (2015 edition) (GB 50010—2010). When there is no frost resistance requirement, the concrete grade shall not be lower than 20 in general; The grade of reinforced concrete shall not be lower than 25.

For the structure of slurry block stone, the strength of the stone soaked in water shall not be lower than 5,000 kN/m^2; the strength grade of cement mortar shall not be lower than 10, and shall not be lower than 20 when frost resistance is required; The strength grade of pointing cement mortar shall not be less than 20.

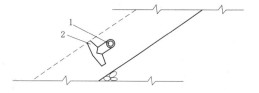

(a) Schematic diagram of random placement of twisted I-shaped blocks

5. Construction of the armour layer

Special construction requirements are added below for several kinds of the mound breakwaters as shown in Fig. 6.2.1.

For the mound breakwater with two-layer twisted I-shaped armour block, when the blocks are placed randomly, the upper layer should have more than 60% of the blocks with the vertical bars below the slope

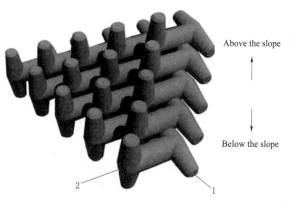

(b) Schematic diagram of regular placement of twisted I-shaped blocks

Fig. 6.2.7 Placement form of twisted I-shaped blocks

1—Horizontal bars; 2—Vertical bars

and the horizontal bars above the slope [Fig. 6.2.7 (a)]. When two-layer twisted I-shaped blocks are placed regularly, all blocks shall remain in the form shown in Fig. 6.2.7 (b). This is because the blocks placed in this way have large anti-tilting moment under the action of waves, which is beneficial to the stability of the armour.

When a layer of accropode blocks is randomly placed, the orientation of adjacent blocks should not be the same, and the hooking of adjacent blocks should be ensured. The blocks at the corner of the slope shoulder should be hooked and embedded. The blocks on the slope shoulder should be in close contact with the breast wall, otherwise corresponding measures should be taken.

Longitudinal and transverse deformation joints and drainage holes shall be set for the slurry block armour. The longitudinal spacing of deformation joints is generally 5 – 10 m, and the transverse spacing is 5 m. The vertical and horizontal spacing of drainage holes is generally 2 – 3 m, and the hole diameter should not be less than 100 mm. The setting of deformation joints can prevent the cracking of the slurry block stone armour due to the uneven settlement of the mound breakwater. The drainage holes are provided to prevent the tidal level on the inner side of the surface protection layer from rising under the action of waves, and then the armour is subjected to excessive water pressure to induce damage due to poor drainage when the waves recede.

The breast wall at the top of the mound breakwater shall be provided with deformation joints. The spacing of the deformation joints shall be determined according to the temperature, structural type, foundation conditions and other factors, which can take the values between 5 m and 15 m, and the width of the deformation joints can take the values between 20 mm and 40 mm.

For the fence panel armour, its longitudinal length $a=1.25H$ and the width along the slope $b=1.0H$. The following are the details of the typical form shown in Fig. 6.2.8:
$a_1=a/15-h/16$; $a_2=a/15+h/16$; $a_3=a/15-h/8$; $a_4=a/15+h/8$; $b_1=0.1b$.

When the cross-section of the riprap block shown in Fig. 6.2.1 (c) is adopted, the ratio of the dimensions of each side of the block is generally $1:1:1.5$.

6. Head and root of the breakwater

Requirements concerning the head and root sections of mound breakwaters are in the following. The length of the head section of the breakwater is generally 15 – 30 m and is usually the slope structure. The length of the head section of the deep-water mound breakwater should not be less than 2 times the height of the embankment body. When there is the requirement to narrow the width of the harbour entrance, the head section of the mound breakwater can be the vertical structure.

The armour blocks on both the inner and outer sides of the head section of the mound breakwater should be heavier than those on the outer slope of the embankment body, or the slope on both sides of the head section should be suitably reduced. This is because the

6.2 Mound breakwaters

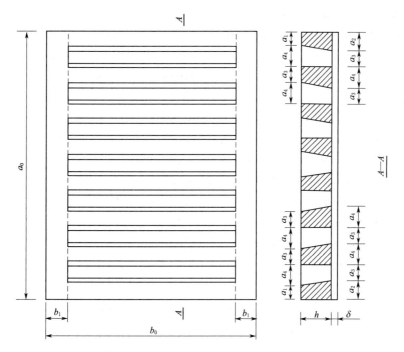

Fig. 6.2.8 Dimensions of fence panels

current induced by wave breaking over the head of the breakwater will push the armour blocks directly outwards from the embankment slope, that is unlike the wave actions on the armour blocks on the embankment body. The weight of the armour block in the head section can be increased by not less than 30% on the basis of the block weight of the embankment body. The block weight of the embankment body as well as the head section in the wave breaking area should be increased by not less than 25% in further. If necessary, it can be determined by model test. The bottom protection structure of the head section should be appropriately strengthened compared with the embankment body. In addition, at the connection between the slope section and the vertical section, the structure shall be strengthened, and the outer slope shall be appropriately reduced.

Generally speaking, after the construction of mound breakwater, the root of the breakwater will be silted up quickly by sediment, and the wave action will gradually weaken, so there is no need to strengthen it. However, if the breakwater is built on the rocky coast, the rocky coast is steep and the water depth at the root of the breakwater is large, the wave energy at the root section may be concentrated due to the reflection of waves on the coast. At this time, strengthening measures should be taken for the root section and the adjacent coastal zone. When the breakwater axis turns outward to form a concave angle, or turns inward to form a sharp convex angle, which will induce significant concentration of wave energy, the strengthening measures should also be taken. And the strengthening range should not be less than one time of the design wavelength value on both sides of the turning position.

6.2.3 Calculation of mound breakwaters

1. Contents of the calculation

The design and calculation contents of mound breakwater and sloping seawall are basically the same. The design of the mound breakwater includes the design of ultimate limit state and the design of serviceability limit state, where the design of the ultimate limit state shall include the calculations and checked of the following contents: ①Strength of fence panels; ② Stable weight of bottom protection block in front of breakwater; ③Strength, anti-sliding and anti-tilting stability of breast wall; ④Overall stability.

The following contents shall be calculated and checked in the design of serviceability limit state: ①Settlement of foundations; ②Width of cracks.

When designing with the ultimate limit state and serviceability limit state of the mound breakwater, the wave force determined by the design wave elements corresponding to the calculated tidal level shall be taken as the standard value. The calculated tidal level in the design condition of ultimate limit state and corresponding combination includes: ① For persistent combination, the design high tidal level, design low tidal level, extreme high water level and extreme low water level shall be adopted for calculation of water level; ② For short-term combination, the design high tidal level and the design low tidal level shall be adopted for calculation of tidal level, or a certain unfavorable tidal level under the temporary state during the construction period; ③ For seismic combination, the calculated tidal level shall be in accordance with the relevant provisions of the seismic standards for water transport projects, and the average tidal level can be taken.

The extreme high tidal level and extreme low tidal level could not be calculated for the action combination of serviceability limit state.

The design wave height in the ultimate limit state and the serviceability limit state of the mound breakwater shall be selected according to the following principles:

(1) In case of extreme high tidal level and design high tidal level under persistent conditions, the corresponding design wave height shall be adopted.

(2) For the design low tidal level in the persistent condition, if there are calculated offshore design waves, the shallow water deformation analysis of the waves is carried out at the design low tidal level to determine the design wave height. If there are only design waves near the building regardless of tidal level statistics, the same design wave height as that at the design high tidal level (but not exceeding the shallow water limit wave height at the low tidal level) shall be taken.

(3) In case of extremely low tidal level under persistent condition, the effect of wave is not considered.

(4) In case of transient condition, the return period of wave height shall be 2 – 5

6.2 Mound breakwaters

years of the check of the construction period of unfinished mound breakwater.

(5) In case of seismic condition, when calculating the overall stability of the mound breakwater, the combination of seismic actions shall be considered without considering the action of waves.

(6) In case of accidental condition, if there are special requirements, the design wave shall be determined according to the corresponding design conditions.

Considering that the design and calculation contents of the mound breakwater are basically the same as those of the sloping seawall, the calculation of seabed scour in front of the mound breakwater is supplemented here, and the principle of seawall calculation in this book can be referred to for other specific calculations.

2. Calculation of seabed scour in front of mound breakwater

Under the action of partial standing waves formed in front of sloping buildings, there are three scour patterns on sandy seabed, including relatively fine sand type, transitional type, and relatively coarse sand types (Fig. 6.2.9). The scour profile characteristics of the relatively fine and relatively coarse sand types are basically the same as those of the corresponding scour patterns in front of the vertical wall under the action of standing waves.

For the transitional scour profile, the positions of scouring valleys and accumulation peaks are deviated from the node and antinode of partial standing waves.

For relatively fine sandy and transitional scour profiles, scour pits occur under the slope surface, so special attention should be paid to the application of protective measures such as the layer of bottom protection blocks in front of the slope.

The same dimensionless discrimination parameter β as that of vertical wall can be used to distinguish different scour patterns of sandy seabed in front of mound breakwater, calculated by Eq. (6.2.2). The maximum bottom flow velocity at the standing wave node can be calculated by Eq. (6.2.3), and the incipient velocity V_{cr} of sand can be calculated by Eq. (6.2.4).

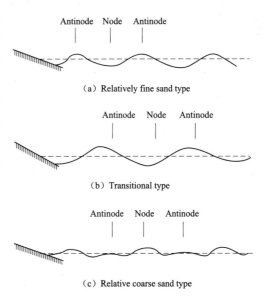

Fig. 6.2.9 Schematic diagram of three scour patterns in front of mound breakwaters

$$\beta = \frac{V_{max} - V_{cr}}{\omega} \tag{6.2.2}$$

$$V_{max} = \frac{2\pi H_{13\%}}{\sqrt{\frac{\pi L}{g}} \sinh \frac{4\pi d}{L}} \tag{6.2.3}$$

$$V_{cr} = 2.4 \Delta^{\frac{2}{3}} D_{50}^{0.433} \overline{T}^{\frac{1}{3}} \tag{6.2.4}$$

$$\Delta = \frac{\gamma_s - \gamma}{\gamma} \tag{6.2.5}$$

Where, β is discriminant parameter of scour patterns; V_{max} is maximum bottom velocity at standing wave node, m/s; V_{cr} is incipient velocity of bottom sediment, m/s, test data shows that Bagnold formula can be used to calculate; ω is sedimentation velocity of sand in still water, m/s; $H_{13\%}$ is wave height with 13% cumulative frequency, m; L is wavelength calculated from average period $\overline{T}$, m; g is gravity acceleration, 9.81 m/s²; d is water depth, m; Δ is relative gravity of sand particles; D_{50} is median grain size of sand, m; $\overline{T}$ is average period of wave, s; γ_s is gravity of sand, kN/m³.

For the scouring pattern in front of the slope, they are the relatively fine sand type for $\beta \geqslant 28$, the relatively coarse sand type for $\beta \leqslant 10$ and the transitional type for $10 < \beta < 28$.

The distance value l is the distance between the first antinode point of partial standing waves in front of the slope and the intersection point of the slope surface and still water surface, which can be calculated as follows.

$$l = \frac{L}{2} - \frac{L}{T \sin\alpha} \left(\frac{R_u}{g}\right)^{\frac{1}{2}} \tag{6.2.6}$$

Where, T is average wave period, s; L is wavelength calculated from average period, m; α is angle of slope, (°); R_u is wave run-up on sloping surface, m.

When the scour pattern is relatively fine sand type or transitional type, the maximum depth Z_{mf} of the scouring valley can be calculated according to Eq. (6.2.7) and Eq. (6.2.8), or it can be determined according to Fig. 6.2.10.

$$Z_{mf} = \frac{0.2 H_{max}}{\left(\sinh \frac{2\pi d}{L}\right)^{1.35}} \tag{6.2.7}$$

$$H_{max} = H_{13\%} + H_R \tag{6.2.8}$$

Where, Z_{mf} is maximum depth of scouring valley, m; H_{max} is wave height at the antinode, m; d is water depth, m; L is wavelength calculated from average period, m; $H_{13\%}$ is wave height with 13% cumulative frequency, m; H_R is reflected wave height, m.

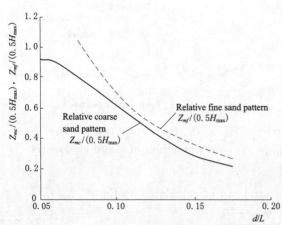

Fig. 6.2.10 Calculation of maximum depth of scouring valley

6.3 Vertical breakwaters

6.3.1 The scale of area of vertical breakwaters

According to the actual situation of existing breakwaters in China, the *Code of Design for Breakwaters and Revetments* (JTS 154—2018) summarizes the determination of cross-section types for five typical vertical breakwaters, including: ①Vertical breakwater of caissons; ②Vertical breakwater of normal placed blocks; ③Vertical breakwater of chamfered blocks; ④Horizontal composite breakwater; ⑤Deep-water vertical breakwater in Fig. 6.1.12.

1. The selecting principles of cross-section types

The wall structure mainly adopts reinforced concrete caissons, concrete blocks or hollow blocks. In Fig. 6.1.12, the (a) & (b), (d) & (c), and (e) of indicate the seaward shapes of the super-structure with the vertical surface, chamfered surface and arc surface respectively.

The cross-section of gravity vertical breakwaters can be selected according to the following principles: reinforced concrete caissons or perforated caissons, or concrete blocks or hollow blocks can be adopted for the construction of the wall of vertical breakwater with rectangular caissons or normal placed blocks. The superstructure can adopt cast-in-place or integral prefabrication and installation concrete structure. The bottom protection structure should be installed in front of the breakwater as needed, as shown in Fig. 6.1.12 (a)-(c). When the breaking wave impact is generated in front of the breakwater, or the existing vertical breakwaters need to be reinforced and repaired, the horizontal composite vertical breakwater with dumping block stones in front of the wall can be adopted, as shown in Fig. 6.1.12 (d). For deep-water vertical breakwaters, the main body of the breakwaters adopts reinforced concrete caissons. The armour protection shoulder and protection slope for foundation bed adopt randomly placed artificial blocks, as shown in Fig. 6.1.12 (e).

2. The crest elevation of the breakwaters

The crest elevation of gravity vertical breakwaters is mainly determined according to the application requirements, combined with the general layout. The following provisions shall be met: For vertical breakwaters allowing a small amount of wave overtopping, it is appropriate to set the crest elevation at 0.6 - 0.7 times the design wave height above the design high tidal level; For vertical breakwaters that basically do not allow overtopping, it is appropriate to set the crest elevation at 1.0 - 1.25 times the design wave height above the design high tidal level; When the superstructure of the vertical breakwaters is a chamfered type, it is appropriate to set the crest elevation at not less than 1.25 times the design wave height above the design high tidal level; For the vertical breakwaters with high protection requirements, the crest elevation is determined by controlling the amount of wave overtopping discharge less than the allowed average wave overtopping discharge according to the protection objects and protection facilities.

According to the relevant model test results, when the top of the vertical breakwater is $(0.6-0.7)H$ above the tidal level, the wave height behind the breakwater after the wave overtopping is about $(0.1-0.2)H$.

3. The crest elevation of wall caissons

The crest elevation of the wall caisson or the topmost layer of the blocks for the vertical breakwater should be not less than 0.3 m above the construction tidal level. And it is generally required to be 0.3 - 0.5 m above the construction tidal level.

4. The width and thickness of the foundation bed

The width of the outer and inner shoulders of the vertical breakwater rubble foundation bed can be 0.6 and 0.4 times the calculated width of the wall body respectively. The width of the wall here refers to the "calculated width". If the wall is widened due to usage requirements, the base shoulder width is not normally increased accordingly. For the width of the outer shoulder of the vertical breakwater with high foundation bed, it can be reduced by the model test results. The side slope gradient of the rubble foundation bed is generally 1 : 2 - 1 : 3 on the outer side and 1 : 1.5 - 1 : 2 on the inner side.

The thickness of the rubble mound foundation of gravity vertical breakwater on non-rock foundation should be determined by calculation. The thickness of the rubble mound foundation should not be less than 1.5 m on cohesive soil foundation and not be less than 0.7 m on sandy soil foundation, and a gravel cushion of 0.3 m should be set under the rubble mound foundation. The thickness of the rubble mound foundation should not be less than 0.5 m on rock foundation. The thickness of the rubble mound foundation should be increased appropriately for deep-water vertical breakwater.

5. The width of the wall

The width of the wall of the vertical breakwater should be determined based on stability calculations.

6.3.2 Structure of vertical breakwaters

1. Rubble mound foundation

Rubble mound foundation of gravity vertical breakwater adopts 10 - 100 kg blocks. When the thickness of foundation is large, 3 m below the top surface of the foundation can adopt hill-excavating stones below 300 kg. The content of particles less than 1 kg in hill-excavating stones shall not exceed 5%. The foundation can adopt layered tamping method or explosive-ramming method for compaction.

2. Bottom protection

Bottom protection block stones in front of gravity vertical breakwaters can construct two layers, and the thickness should not be less than 0.5 m. When bottom protection

6.3 Vertical breakwaters

block stones' weight is greater than 100 kg or with sandy seabed, gravel mattress and geotextile mattress no less than 0.3 m should be set under the layer of bottom protection block stones.

3. Superstructure

The superstructure of gravity vertical breakwaters should have sufficient stiffness and good integrity, and be firmly connected with the wall structure. The thickness of the superstructure should not be less than 1.0 m, and the superstructure of the caisson body should be embedded in the caisson, and the embedded depth should not be less than 0.3 m. According to the wave and construction conditions, the upper part of the caisson can be considered to be capped with concrete. When the superstructure adopts chamfered type, the slope angle adopts 25°–30°.

4. The weight of vertical breakwaters block structure

When concrete blocks are used for the wall, their weight is generally determined according to the capacity of lifting equipment, and the weight of concrete blocks should meet the stability requirements, while not less than the values in Table 6.3.1. When the weight of the square block can not meet the stability requirements, concrete can be poured in the middle of reserved holes. The types of concrete blocks for breakwater body should be reduced, the ratio of the length of long sides of the blocks to the height should not be greater than 3.0. The ratio of the length of short sides to the height should not be less than 1.0 (should not be less than 0.8 for individual blocks).

Table 6.3.1　　　　　　　　**Minimum weight of vertical breakwaters blocks**

Design wave height $H_{1\%}/m$	Weight of blocks/t	Design wave height $H_{1\%}/m$	Weight of blocks/t
2.6 – 3.5	30	5.6 – 6.0	60
3.6 – 4.5	40	6.1 – 6.5	80
4.6 – 5.5	50	6.6 – 7.0	100

Vertical breakwater of normal placed blocks should reserve through holes for concrete pouring when the stability of the upper layer blocks is insufficient or when the number of layers of the blocks wall exceeds 7 layers. The overall reinforcement measures should be taken for vertical breakwaters of normal placed blocks with seismic requirements.

For the vertical breakwaters in the frozen area, the waterfront surface in the tidal level changing area should take measures to enhance durability, such as increasing the protective layer of reinforcement or inlaying the surface with granite.

5. Vertical joints and deformation joints

The width of vertical joints between blocks in vertical breakwater of normal placed blocks can be 20 mm, and the vertical joints should be staggered with each other. The spacing of the staggered joints should not be less than the values in Table 6.3.2. Under special circumstances, the spacing of the staggered joints can be reduced to 400 mm in the longitudinal section or on each layer plane, but the number of them should not exceed 10% of the corresponding total number of joints.

Table 6.3.2 Spacing of the staggered joints

Staggered joints position	Weight of blocks	
	Less than 40 t	Over 40 t
In cross-section	0.8 m	0.9 m
In longitudinal section or plane	0.5 m	0.6 m

Note: Under special circumstances, the spacing of the staggered joints can be reduced to 0.4 m in the longitudinal section or on each layer plane, but the number of them should not exceed 10% of the corresponding total number of joints.

The width of the vertical joints between caissons in vertical breakwater of caissons should adopt 4‰ of the caisson height, but should not be less than 50 mm, and should not be greater than 100 mm. Caissons can use flush or butt joints. When the butt joints are applied, the corresponding cavity width can adopt 0.3 – 0.8 m.

Deformation joints should be set along the length of the vertical breakwater, and the deformation joints should be made into continuous-seam with width of 20 – 50 mm. The spacing of the deformation joints should be determined according to the temperature, structure type, foundation conditions and foundation bed thickness, etc. The block structure is generally 10 – 30 m. The vertical joints between caissons are generally used as deformation joints for caisson structures. Deformation joints should be set at the following locations: the location for changing the structure types, the location for suddenly changing of wall height or foundation thickness, the location where soil quality of the foundation varies greatly.

6. Head and root of vertical breakwaters

As mentioned above, the load combination for the design of the head of breakwaters is the same as that of the embankment body. Therefore, the width of the head of vertical breakwaters is generally the same as the body section, unless it is required to increase the width due to the setting of beacons or lighthouses at the head. In the plane, the two corners of the head can be filet or cut (Fig. 6.3.1) for smoother wave and current bypassing.

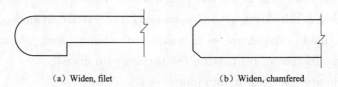

(a) Widen, filet   (b) Widen, chamfered

Fig. 6.3.1 Plane form of the head

In the head section of the vertical breakwater, the waves and current are more turbulent, so the inner and outer base shoulders should be strengthened. According to the water depth and wave elements in front of the breakwater, base shoulders in the head section should be strengthened with different types of artificial blocks such as four-legged hollow blocks. The bottom protection structure of the head section should also be strengthened appropriately compared with the body section, and the bottom protection block layer should be extended from the outer side to the inner side of the embankment. The slopes of the inner and outer sides of the rubble mound foundation of the head section should be smaller than the slope of the sea side of the body section. For the connection location between the vertical section and the slope section, the structure should be strengthened. The length of the head section of the vertical breakwater that needs to be reinforced by the above-mentioned measures should be 1.5 – 2.0 times the width of the embankment and should not be less than 1.0 times the height of the embankment.

Due to the shallow water depth around root section, it is difficult for construction vessels to enter the root section of the embankment, and the slope type is often more economical. So even if the body section of the embankment is vertical, the root section of the embankment is generally the slope type structure. And this should comply with the relevant provisions of the slope structure.

6.3.3 Calculation of vertical breakwaters

1. Calculation contents

The calculations for gravity vertical breakwaters' stability against overturning and stability against sliding are basically the same as those for gravity quay and steep slope seawall. Only the principles of load combination and the items that need to be calculated are specified here, without discussing the specific calculation methods.

For the load combination of vertical breakwaters, it can be generally divided into design combination and verification combination.

For the design combination, when the design high tidal level is used for the calculation, the design wave height is used for the wave height. When the design low tidal level is used for the calculation, the determination of wave height is divided into the following two cases: when there are calculated offshore design waves, the design low tidal level can be taken for wave refraction analysis to obtain the design wave height in front of the breakwater; when there are only design waves counted without the distinction of tidal levels near the proposed breakwater, the same design wave height as that of the design high tidal level can be taken, but it should not exceed the limit wave height of shallow water at the low tidal level.

If the wave motion in front of the breakwater is standing wave at the design high tidal level, and wave breaking phenomenon (broken waves or breaking waves) occurs at the design low tidal level, the tidal level situation that may generate the maximum wave force between the design low tidal level and the design high tidal level should be calculated.

For the verification combination, when the verification high tidal level is used for the calculation, the design wave height is generally used for the wave height. If there are calculated offshore design waves, the verification high tidal level can be taken for wave refraction analysis to obtain the wave height in front of the breakwater. When the verification low tidal level is used for the calculation, the effect of waves is not considered. This is because the calibration low tidal level is generally induced by offshore wind in form of the wave level reduction, so there will be no shoreward waves.

The code of breakwaters specifies that for gravity vertical breakwaters, three design conditions and corresponding combinations should be considered when the design is according to ultimate limit state.

(1) The persistent condition should be considered, which is the design combination and verification combination about the vertical breakwater mentioned above. The *Code of Design for Breakwaters and Revetments* (JTS 154—2018) uses extreme high and low tidal levels instead of the verification high and low tidal levels.

(2) The transient condition should be considered. For the verification in construction period of unfinished gravity vertical breakwaters, the tidal level may adopt the design high and low tidal level, and the return period of wave height may adopt 5 – 10 years.

(3) In the calculation of the foundation bearing capacity and overall stability of the gravity vertical breakwater, the accidental combination of seismic action should be considered.

The calculation method should be consistent with the relevant anti-seismic provisions of the water transport engineering, but should not consider the combination of wave and seismic action. This is because when the seismic load acts alone (not combined with wave force), the value is much smaller than the wave force in design combination. Calculations have been made for three typical vertical breakwaters in China, and the results show that even if the seismic intensity of 9 degree is considered, the seismic force in the horizontal direction (self-weight inertia force plus ground seismic hydrodynamic pressure) is less than half of the design wave force. On the other hand, from the actual situation domestic and international, there is no combination of seismic load and wave force generated by the design waves. Even in countries with frequent earthquakes like Japan, this combination is not considered when designing the vertical breakwaters.

When calculating the stability of the vertical breakwater, it is common to disregard the combination of waves on the land side of the breakwater and waves on the outer side, which means the water surface on the land side of the breakwater is calculated as still water surface. This is because the waves on the inner side of the breakwater are basically propagating along the embankment and there is no possibility of wave propagating towards the embankment body. When the crest elevation of the breakwater is relatively low, although the inner side of the harbor will have waves induced by the water overtopping the breakwater, the wave force on the outer side of the embankment will also be reduced by

6.3 Vertical breakwaters

the waves overtopping. Therefore, for the calculation, it can be assumed that the water surface on the inner side of the breakwater is the still water surface.

For head of the breakwater, the waves near the embankment head are more turbulent according to the field observation, and there is a possibility of water surface reduction on the inner side. However, due to the bypass flow at the head of the breakwater, it is impossible to induce total reflection for a distance outside the head of the breakwater, so its wave pressure will be smaller than the standing wave pressure. The conclusion that the wave pressure at the head of the breakwater is smaller than that at the embankment body is proved by some theoretical analysis, model tests results and engineering practice. Therefore, for the stability calculation of the head of the breakwater, it is acceptable to use the combination of wave pressure of the design wave height outside the breakwater and still water inside the breakwater.

The following is a list of the main items that should be calculated for designing the vertical breakwaters.

1) Stability against overturning of each horizontal joints and tooth seam along the bottom and body of the breakwater.

2) Stability against sliding of each horizontal joints along the bottom and body of the breakwater.

3) Stability against sliding along the bed.

4) Bed and foundation stresses.

5) Overall stability.

6) Foundation settlement.

7) Stability weight of the shoulder protection block of the bed and the bottom block in front of the breakwater.

For reinforced concrete caisson structure, the following items should also be calculated.

1) Draught, freeboard height and floating stability of caissons.

2) Strength and cracking resistance of the outer wall, the partition wall, the bottom plate and the cantilever of the bottom plate of caissons.

When calculating the outer wall, the combination of the lateral pressure of sand, stone, or other fillers in the caissons, and the wave trough pressure outside the caissons should be considered. For the partition wall as the outer wall support, it should be calculated and checked for the cross-section under tensile conditions.

The wave buoyancy uplift pressure in each seam of the embankment body needs to be taken into consideration when calculating the stability against overturning along each horizontal seam and tooth seam of the embankment body and the stability against sliding along each horizontal seam of the embankment body. The wave buoyancy uplift pressure p in the square seam beneath the water surface under the action of wave crest can be distributed as the triangle pattern along the width of the embankment body, and its maximum pressure is equal to the wave lateral pressure at the same elevation. For the wave lateral pressure in the "bc" seam (refers to the vertical seam between b, c) in Fig. 6.3.2, the same wave buoyancy uplift pressure could be adopted for b, c points, the pressure between b, c points can be provided as the linear distribution.

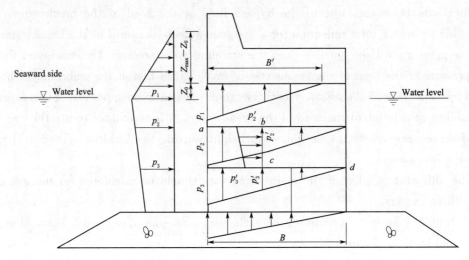

Fig. 6.3.2 Wave pressure in the square seam of the embankment

The effective action width B' of the wave buoyancy uplift force acting in the integral or assembled integral superstructure under the ground can be calculated according to the following formula.

$$B' = B \frac{Z_{max} - Z_0}{Z_{max}} \qquad (6.3.1)$$

Where, Z_{max} is the height of the wave crest above the still tidal level, m; Z_0 is the height of the bottom surface of the superstructure above the still tidal level, m.

The wave buoyancy uplift pressure p' along B' can be as a triangular distribution, and its maximum pressure is equal to the wave lateral pressure at the same elevation.

The wave buoyancy of the square seam beneath the water surface under the action of wave trough can be calculated according to the same principle as that under the action of wave crest.

The overall stability of vertical breakwater built on a non-rock foundation is generally calculated using the circular slip surface method. When there exists the soft soil interlayer, the non-circular slip surface method should be used. In the verification calculation of circular slip surface of vertical breakwater, when the center of the slip circle is higher than the action point of the wave lateral pressure under the wave crest action, the circular arc slipping towards the port shall be drawn. When the center of the slip circle is lower than the action point of the wave pressure, the circular arc slipping out of the port shall be drawn. Under the action of wave trough, drawing directions of the circular arc are the opposite, which means, when the center of the circular arc is higher than the action point of the wave pressure, the circular arc slipping out of the port shall be drawn; when the center of the circular arc is lower than the action point of the wave pressure, the circular arc slipping towards the port shall be drawn. The slipping circular arc generally passes through the front toe or heel of the wall. In addition, if scour holes are allowed on the seabed in front of the breakwater for the design, the adverse effects of scour holes shall be considered when checking the circular arc slipping out of the port.

When calculating the foundation settlement of vertical breakwaters, the settlement value caused by wave lateral pressure can usually be ignored. This is because the horizontal force has little effect on the settlement, and the wave force is not the long-term load. The allowable average settlement of vertical breakwaters is generally taken as 35 cm for caisson structure and 30 cm for square structure, and the reserved settlement of the crest elevation of vertical breakwaters can be determined according to the foundation and construction conditions.

2. Calculation of the stability weight of the base shoulder and slope blocks

Two calculation methods for determining the stability weight of the base shoulder and slope blocks of the rubble mound foundation are described below.

In the appendix of *Code of Design for Breakwaters and Revetments* (JTS 154—2018), the block weight calculation diagram drawn according to the regular wave test results of Canada is given (Fig. 6.3.3).

For the stability calculation of the rubble mound foundation and the bottom protection blocks of the vertical breakwater, the wave height H should adopt accumulative frequency of 5% of the wave height $H_{5\%}$.

On the basis of the water depth d in front of the embankment and the water depth d_2 on the protective layer of the foundation bed, from d_2/d on the horizontal axis of the right half of Fig. 6.3.3, draw an upward vertical line to intersect the corresponding d/L, then draw a horizontal line from the intersection point to the left to intersect with the contour line of H and then draw the downward vertical line from this intersection point, the block stability weight W is obtained on the horizontal axis of the left half of figure (in kg or t). In addition, there is alternative way to draw a horizontal line from the first intersection point to the left to intersect with the vertical axis to obtain the coefficient K and then substitute this coefficient into the formula $W=KH^3$.

If the block is placed, the weight can be approximately 0.6 times of the weight of the riprap W.

When the slope of the rubble mound foundation is 1:1.5, the block weight can be approximately 1.33 times of W in Fig. 6.3.3.

According to the engineering practical results, the above methods sometimes give excessive block weights. For this reason, the results obtained using irregular wave tests

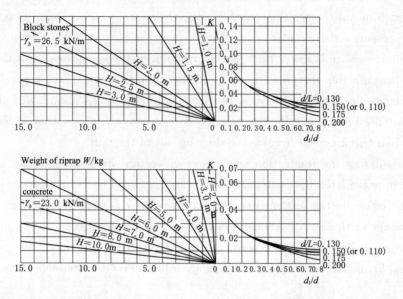

Note: 1. d_1 is water depth on top of the foundation bed, m; d is water depth in front of the embankment, m; H is design wave height, m, adopting wave height with accumulative frequency of 5% of the wave height $H_{5\%}$; L is calculation wavelength, m.
2. The coefficient K can be found from d_1/d and d/L in the right half of the figure, and the block stability weight $W = KH^3$, and the corresponding block stability weight can also be found from the coefficient K and the wave height H in the left half of the figure.
3. If the block is placed, the weight of the block can be approximately by 0.6 times of the weight of the riprap. When the slope is 1:1.5, the weight of the block can be approximately 1.33 times of the value in the figure.

Fig. 6.3.3 Calculation of the stability weight of the base shoulder and slope blocks of the rubble mound foundation

by the Port and Harbor Research Institute of the Ministry of Transport of Japan are presented below. The stability weight of the armour block of the rubble mound foundation is

$$W = \frac{\gamma_b H_S^3}{N_S^3 (S_b - 1)^3} \qquad (6.3.2)$$

Where, γ_b and S_b are the weight and specific gravity of the block; N_S is the stability number, which depends on the ratio of the water depth d on the foundation bed to the significant wave height H_S, d/H_S, as well as the type of the armour and the parameter K. K represents the comprehensive influence of the relative water depth and the relative distance from the standing wall on the maximum horizontal bottom velocity, which can be calculated using Eq. (6.3.3).

$$K = \frac{\dfrac{4\pi d_1}{L_1}}{\sinh \dfrac{4\pi d_1}{L_1}} \sin^2\left(\frac{2\pi b_m}{L_1}\right) \qquad (6.3.3)$$

Where, L_1 is the wavelength corresponding to the water depth d, which can be calculated by the dispersion relation of the significant wave period T_S (replacing $\overline{T}$), m; b_m is width of the base shoulder, m.

Fig. 6.3.4 represents the stability number N_S derived from the irregular wave test of

6.3 Vertical breakwaters

the two layers dumping block stones armour at slope 1 : 2, which is determined by d_1/H_S and K.

It should be noted that, the use of H_S in Eq. (6.3.2) does not imply that H_S is the equivalent wave height. In fact, in this test made by Port and Harbor Technology Research Institute of the Ministry of Transport of Japan, the comparison of the regular wave test with the irregular wave test indicates that the equivalent wave height is about 1.371.

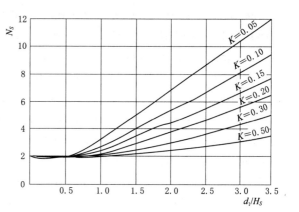

Fig. 6.3.4 N_S diagram of armour blocks of the rubble mound foundation

The relationship between the stability weight of the bottom protection blocks in front of the vertical breakwater and the maximum wave-induced bottom velocity in front of the breakwater is the same as that described in the previous section. The maximum wave-induced bottom velocity when the wave field state in front of the breakwater is standing wave is

$$V_{max} = \frac{2\pi H}{\sqrt{\frac{\pi L}{g} \sinh \frac{4\pi d}{L}}} \qquad (6.3.4)$$

Where, H adopts $H_{5\%}$, Eq. (6.3.4) is consistent with the Eq. (6.2.3) in the specification for breakwaters. The difference is that, in calculating the sand bottom scour in front of the breakwater, H in the formula adopts $H_{13\%}$ according to the specification for breakwaters.

When the wave field state in front of the breakwater is broken waves, the maximum wave bottom flow velocity can be calculated by Eq. (6.3.5).

$$V_{max} = 0.33\sqrt{g(H+d)} \qquad (6.3.5)$$

When the wave field state in front of the breakwater is breaking waves, the maximum wave bottom flow velocity can be calculated approximately by Eq. (6.3.4), but H adopts $H_{5\%}$. The width of the vertical breakwater's bottom protection blocks layer is generally 0.25 times the design wavelength. If scour holes are allowed on the seabed in front of the breakwater in the design, the location and scale of the scour holes can be determined as described previously and the adverse effects of the scour holes can be considered in the overall stability calculation as described in this section.

The thickness of bottom protection layer is generally the thickness of 1 – 2 layers of blocks, and should not be less than 0.5 m. When the weight of the bottom protection blocks is more than 100 kg, the bedding layer with small blocks should be set under it.

The thickness of the rubble mound foundation of vertical breakwater is generally determined by calculation, but should not be less than 1.0 m on non-rock foundation. The rubble mound foundation generally adopt 10 - 100 kg of stones and is tamped with heavy hammer. However, for high bed vertical breakwater, it can be determined whether to tamp according to the foundation conditions, usage requirements and working conditions.

3. Wave scouring of sand bottom in front of vertical breakwaters

The scouring of the sand bottom in front of vertical breakwaters by the waves will affect the overall stability of the breakwater.

The sand bottom in front of the vertical breakwater could have two basic scour patterns under the action of standing waves. The results of movable-bed bed model tests show that the occurrence of these two scour patterns depends on the sediment grain size and the wave element.

When the particles composing the sand bottom are fine, the sediment movement state is dominated by suspended load, and sand particles will be transported according to the circulation pattern of mass transport in standing wave, so the sediment movement direction at the bottom will be from the nodes (in front of the breakwater $n\frac{L}{4}$, $n=1,3,5,\cdots$) to the antinodes (in front of the breakwater $m\frac{L}{4}$, $m=0,2,4,6,\cdots$), that is the same direction as the mass transport flow there. As a result, the scour pattern of fine grain seabed is that scour occurs at the nodes of standing wave, while accumulation occurs near the antinodes [Fig. 6.3.5 (a)].

When the particles constituting the sand bottom are relatively coarse, sediment movement is dominated by bed load. The sand bottom is scoured in the middle of the nodes and the antinodes, and the accumulation occurs at the nodes [Fig. 6.3.5 (b)].

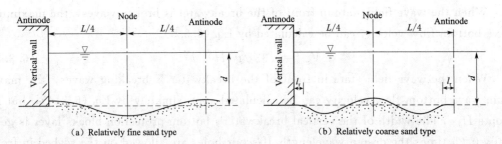

Fig. 6.3.5 Schematic diagram of the submarine scour pattern in front of the vertical breakwater

l_r is scour profile parameter, m; L is calculated wavelength, m; d is water depth in front of the breakwater, m

The analysis of the experimental data shows that, the same as scour analysis in front of mound breakwaters in Section II, dimensionless discriminant parameters β can distinguish the two basic scour patterns mentioned above. When $\beta \geqslant 16.5$, it is the scour pattern of relatively fine particles; when $\beta < 16.5$, it is the scour pattern of relatively coarse par-

6.3 Vertical breakwaters

ticles.

For most of the actual breakwater projects, relatively fine-grain scour patterns will mainly occur in front of the breakwater. For the relatively fine-grain scour pattern, the equilibrium scour profile can be approximately described by the trochoid line.

$$x_t = \frac{L}{4\pi}\theta + R\sin\theta \tag{6.3.6}$$

$$z_t = -R\cos\theta \tag{6.3.7}$$

Where, x_t and z_t are the horizontal and vertical coordinate values of the scour profile; θ is the calculated angle (rad) from 0 to 2π; x_t measured from the node and z_t measured from the height z_0 on the original sand bottom surface, upward positive.

$$R = 1 - \frac{\sqrt{1 - \frac{8\pi Z_{mf}}{L}}}{4\pi/L} \tag{6.3.8}$$

$$z_0 = R - Z_{mf} \tag{6.3.9}$$

Where, R is the parameter of the trochoid profile, m; Z_{mf} is the maximum depth of the scour valley when the scour profile reaches equilibrium, which can be calculated by the following equation.

$$Z_{mf} = \frac{0.4H}{\left(\sinh\frac{2\pi d}{L}\right)^{1.35}} \tag{6.3.10}$$

According to the comparison results of irregular wave and regular wave tests, the H in the above equation can adopt the significant wave height H_S; the wavelength L can be calculated by the average period $\overline{T}$.

When there is a riprap protective layer at the upstream of the seabed in front of the breakwater, the shape of the scour profile will be different from that without the protective layer, as shown in Fig. 6.3.6.

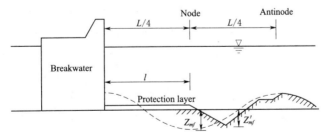

Fig. 6.3.6 Effect of protective layer on sand bottom scouring

The influence of the protective layer on the sand bottom is mainly limited to $L/2$ in front of the breakwater. When Z'_{mf} represents the final scour depth of the first scour valley in front of the breakwater with the protective layer, Z'_{mf} decreases with the increase of the width of the protective layer 1 in front of the breakwater. The width of the scour valley also decreases with the increase of l. While the distance of the first scour valley from the breakwater increases with the increase of l.

For the scour pattern of relatively fine particles, when $l=3L/8$, there will be basically no scouring within $L/2$ in front of the breakwater. For the scour form of relatively coarse particles, when $l=0.15L$, the first scouring pit in front of the embankment can be generally prevented.

6.4 Mixed breakwaters

6.4.1 Structural type of mixed breakwaters

The type with a sloping cover prism in front of the upright wall is referred to as a mixed breakwater. In principle, various structural types of vertical breakwater and mound breakwater are suitable for forming mixed breakwater.

The East Breakwater of Yantai Gang was built from 1915 to 1920. The structure of the breakwater was originally the inclined block vertical breakwater (the inclined block structure is rarely used now). The blocks and superstructure were made of slurry block stones. The cross-section of the breakwater was shown in Fig. 6.1.15. The head of the breakwater was made of reinforced concrete caissons.

This breakwater encountered a large storm during the construction process, which caused serious damage to the 80 m-long breakwater's body that had been placed. A number of blocks on the top were knocked into the sea by the wind waves, and more slurry block stones were knocked so that blocks fell off and mortar joints cracked. Three columns of embankment blocks were all displaced and inclined, so that gaps were induced between the columns of blocks. After that, except for the replacement of severely damaged blocks, they were generally filled with bagged concrete. The blocks inclined outwards on the sea side of the embankment were not relocated, but a series of special foot protection blocks weighing about 40 t were added in front of the breakwater (Fig. 6.1.15). But this remedy was not complete and appropriate, and instead contributed to the wave breaking on the foot protection blocks. Therefore, the damage of this section of the breakwater had always been more serious than that of the other sections, and many large holes were formed on the double-layer blocks on the top of sea side embankment by incident waves. The holes' sealing method with bagged concrete and filling method by pouring concrete in the holes were applied to repair, but it also failed. After 1952, block stone prisms were placed in the sea side of this section for the breakwater. The top elevation of the prisms was about the same as that of the low tidal level. The block stones weigh about 150 – 300 kg, and the 7 t blocks were pressed on the top of the prisms. However, this prismatic body with relatively low elevation in front of the breakwater intensified the wave breaking and its effect on the breakwater body, resulting in more large holes.

6.4 Mixed breakwaters

In order to completely repair and strengthen this breakwater, hydraulic physical model experiments were carried out in the 1960s. The results showed that larger prisms with its top elevation higher than the design high tidal level needed to be cast in front of the breakwater. For the protection structure of prism, many schemes such as tetrapod, twisted I-shaped block, and accropode block had been tested. The final cross-section of the mixed breakwater used is shown in Fig. 6.1.15, with the slope armour contained two layers of I-shaped blocks with 2.5 t weight.

Gansbee Fishing Port in South Africa is located in the open rocky coast. At first, it was determined to build the vertical breakwater because the inner side of the breakwater was used as wharf. According to the investigation, the design wave height of 6 m was adopted. The width of the vertical wall was 7.3 m. The reinforced concrete hollow blocks of mesh shape were adopted. The blocks' length was equal to the wall width, the width (along the length of the breakwater) was 3 m, and the wall thickness was 30 – 40 cm. Each hollow block was about 3 m in height. The hollow blocks were stacked on the gravel foundation bed, and then the hollow part was filled with concrete. Holes were bored in the cast-in-place concrete until 12 – 15 m below the seabed, and then steel wire anchor bowls were set in the boreholes. The cross-section of this breakwater, including the prisms in front of the breakwater added later, is shown in Fig. 6.4.1.

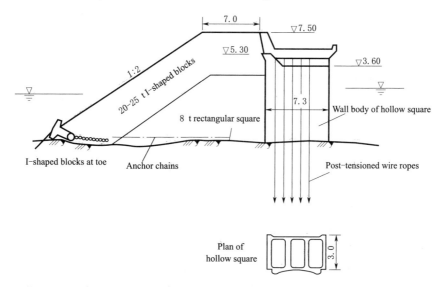

Fig. 6.4.1　Reinforcement of vertical breakwater with prism in front (unit: m)

The wave observation after the breakwater construction showed that the design wave height could reach 8 m. In addition, a large amount of wave overtopping often made it difficult to use the berth on the inner side of the breakwater. Moreover, ultrasonic detection showed that the steel cable anchor bowls had been eroded after being used for about 10 years, especially in the areas between the hollow blocks of each layer and between the hol-

237

low blocks and the seabed. For these reasons, the original cross-section of the vertical breakwater was finally modified and reinforced. The top elevation of wave retaining wall was increased from 5.3 m to 7.5 m. The concrete blocks of 8 t weight were cast in front of the embankment, and the sea side of the breakwater was protected with 20 – 25 t I-shaped blocks. Since the rock seabed was relatively smooth, the model experiments displayed that the I-shaped blocks at the foot of the slope were easy to slide outward, so these blocks were anchored in the prism with heavy chains and anchor blocks (Fig. 6.4.1).

In the case of large wave height, in order to reduce the wave force acting on the vertical wall, it could also be designed as the mixed breakwater directly. Fig. 6.1.16 shows the cross-section of the West Breakwater of Rumoi Port in Japan. The maximum local wave height of 50-year return period was up to 13 m, so the mixed embankment was designed. The vertical wall was composed of caissons, and the slope in front of the breakwater was covered with 50 t tetrapod. Compared with the vertical breakwater, the prism in front of the breakwater can effectively reduce the wave force, while compared with the general mound breakwater, the wave overtopping can be reduced due to the wider width of the entire top of the breakwater. In addition, the vertical wall prevented the wave from penetrating into the harbor.

6.4.2 Wave force of mixed breakwaters

For the mixed breakwater with slope protection prism in front of the vertical wall, the wave force acting on the wall can also be calculated by Goda formula in principle, and takes $\lambda_1 = \lambda_3 = 0.8 - 1.0$ and $\lambda_2 = 0$. The ranges of λ_1 and λ_3 are related to the ratio of the prism width to the wavelength. Generally speaking, when the vertical wall is fully covered by the wave dissipation blocks (Fig. 6.4.1), the values of λ_1 and λ_3 can be taken as 0.8.

6.5 Floating breakwaters

Up to now, there are no less than dozens of types of floating breakwaters proposed by various countries, but most of them are still in the laboratory test stage or the prototype experimental stage, and relatively mature and practical types are still rare. Various types of floating tanks are mostly used, which can be welded from steel plates, used as barges, or as reinforced concrete mooring barges. The rectangular pontoon floating breakwater was built in 1976 in Fukuyama Prefecture, Japan, and the local water depth was from −5.5 to 16.0 m. The high tidal level was 3.84 m and the low tidal level was ±0.0 m. The maximum wave height with 30 – year return period was 2.3 m, and the corresponding wave period was 3.5 s. The total length of the floating breakwater was 275 m, consisting of two 60 m floating tank sections and two 70 m floating tank sections. The floating tank was 10 m in width, 3 m in depth and 2 m of draught. The longitudinal spacing between the neighboring floating tanks was 5 m. Each buoyancy tank had 6 anchors. Fig. 6.5.1

shows the schematic diagram of short floating tank type breakwater.

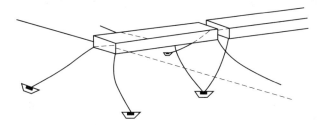

Fig. 6.5.1   Pontoon floating breakwater

It is generally required that the natural frequency of the floating breakwater is lower than the wave frequency, so that more wave energy could be reflected in order to reduce the wave energy transmitted into the port water area. When the floating body has a large mass or a large moment of inertia, its natural frequency is relatively low. After ballasting, the pontoon breakwater generally has a large mass and relatively small movement. The design principle of the A-shaped frame floating tank shown in Fig. 6.5.2 is to make the floating body have a large moment of inertia. A vertical wave baffle is set in the middle of this floating breakwater, and there are two cylinders on each side. The plates and cylinders are connected by an inverted A-shaped frame. The cylinders

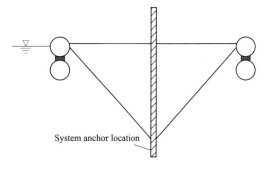

Fig. 6.5.2   A-frame floating breakwater

below on both sides are perforated, with the hole fraction of 50%. The A-shaped frame floating breakwater was proposed by Canada and has been put into use.

In addition to steel, wood, reinforced concrete and other materials, the floating breakwater can also be made of water filled nylon bags, glass fiber reinforced plastics or waste automobile tires. In USA, there are a large number of waste automobile tires to be treated, so several floating breakwaters composed of waste tires have been developed. And there are several examples of this kind of the floating breakwater in USA. It is mainly used in the semi-sheltered area with the distance between the opposite bank less than about 10 km. The effective wave height should not be greater than about 1.0 m. The width of the floating raft usually needs to be about 1 time of the wavelength to dissipate the waves effectively.

6.6   Optimal design of breakwaters

6.6.1   Optimal design of mound breakwaters

When the conventional method is adopted for the breakwater design, the standard design wave return period is generally determined at first, while the wave return period is not specified in advance when the breakwater is optimally designed. For different waves whose return periods vary within the certain ranges, each breakwater cross-section is designed according to the same structural type and design principles, and the costs are determined separately. For each cross-section, when the wave exceeds its adopted value, the expected damage can be determined by model tests or calculations, and the cost required for maintenance can be estimated. The best cross-section of the breakwater is the one with the minimum total investment including construction and maintenance costs. In fact, the repair cost above is only the direct loss caused by the breakwater damage. If the damage of the breakwater affects the port operation, or even causes damage to the quay or the revetment and other buildings in the port, these indirect losses should also be considered. Obviously, the optimal design of breakwaters discussed here is different from the usual concept of optimal design of structural systems under certain loads. Of course, there is also the optimal design of breakwater cross-sections under the action of deterministic wave elements, and this case is not discussed here.

The following example of riprap mound breakwater will be used to specify the theory and methods used in the optimal design of the mound breakwater. The empirical cumulative frequency curve of the significant wave height H_S is obtained from the statistical analysis of wave observation data at certain port. Assuming that each observed wave is generated by one "storm", there are $M = 1,460$ independent "storms" of different sizes per year, since the waves are observed four times a day. The cumulative frequency $P = 1/1,460 = 6.85 \times 10^{-4}$ corresponds to the wave height with one-year return period. The values of H_S and P in columns 1 and 2 of Table 6.6.1 are derived from the empirical cumulative frequency curve. The wave frequency for the wave height greater than H_{S1} and less than H_{S2} is $\Delta P = P_1 - P_2$. P_1 and P_2 are the cumulative frequency values corresponding to H_{S1} and H_{S2}, respectively. The wave height in this range can be represented by $H'_S = (H_{S1} + H_{S2})/2$. Those are columns 3 and 4 of the table.

Five riprap mound breakwater cross-sections were designed using wave heights of 5.7 m, 6.75 m, 7.0 m, 7.25 m, and 7.5 m, following the same design principles. For simplified expressions, only the data for wave heights of 5.7 m and 7.0 m are given in Table 6.6.1. The weight W of armour blocks for mound breakwater is calculated using Hudson formula, but K_D and n are renamed as damage factor and damage rate in this section.

If the wave height and damage factor for no damage (equivalent to an allowable damage rate of $n\%$) are H_{S0} and K_{D0}, the damage factor ratio can be obtained.

$$C = \frac{K_D}{K_{D0}} = \left(\frac{H'_S}{H_{S0}}\right)^3 \tag{6.6.1}$$

Fig. 6.6.1 shows the damage rate of the riprap embankment with different values of C for a small amount of overtopping waves allowed from the experiments. C is obtained

6.6 Optimal design of breakwaters

Table 6.6.1   Mound breakwater optimization calculation table

H_S/m	Wave conditions				Design wave height 5.7 m				Design wave height 7.0 m			
	P	H_S'	ΔP	P_D	C	Damage rate/%	Maintenance costs	Annual maintenance costs	C	Damage rate/%	Maintenance costs	Annual maintenance costs
1	2	3	4	5	6	7	8	9	6'	7'	8'	9'
5.5	8.50×10^{-4}	5.8	4.33×10^{-4}	0.46900	1.05	8.0	2,218	1,039				
6.0	4.17×10^{-4}	6.3	2.17×10^{-4}	0.27200	1.35	18.0	4,991	1,355				
6.5	2.00×10^{-4}	6.7	1.50×10^{-4}	0.19700	1.62	25.0	8,996	1,769				
6.9	5.00×10^{-5}	7.0	3.74×10^{-5}	0.05310	1.85	31.0	11,155	593	1.00	2.5	1,095	58
7.1	1.26×10^{-6}	7.2	9.60×10^{-6}	0.01390	2.02	36.0	12,954	180	1.09	10.0	4,380	61
7.3	3.00×10^{-6}	7.5	3.00×10^{-6}	0.00437	2.28	43.0	21,250	93	1.23	15.0	6,570	29
7.7	0					Total		5,029		Total		148

Note: All costs units in the table are in Yuan/m.

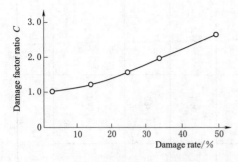

Fig. 6.6.1 Damage rate at different C values

from the design wave height (H_{s_0}) and each level of wave height in Table 6.6.1 (H'_s), and the corresponding damage rate could be found on Fig. 6.6.1. This is the 6th (or $6'$) and 7th (or $7'$) columns of Table 6.6.1. With the damage rate, the quantity of maintenance work can be calculated, and the maintenance cost per time can be calculated according to the unit price (column 8 or $8'$ in Table 6.6.1).

In order to obtain the annual maintenance cost corresponding to a certain level of wave height, it is necessary to know the annual probability of damage P_D corresponding to a certain level of wave height at first.

If the damage is repaired immediately for each time, P_D is the annual probability of damage occurrence corresponding to a certain level of wave height.

$$P_D = M\Delta P \tag{6.6.2}$$

The second situation is to repair once a year. Due to the ΔP as the occurrence frequency of a certain level of wave height in a "storm" duration (6 h), $1-\Delta P$ is the probability of its non-occurrence, and $(1-\Delta P)^M$ is the probability of its non-occurrence in M times "storms", that is, in a year. Then the occurrence probability of a certain level of wave height in one year is

$$P_D = 1 - (1-\Delta P)^M \tag{6.6.3}$$

Similarly, if the restoration is once every m years, the corresponding occurrence probability is

$$P_D = 1 - (1-\Delta P)^{mM} \tag{6.6.4}$$

In this example, it is assumed to be repaired once a year, column 5 in Table 6.6.1 is the product of the calculation result P_D according to Eq. (6.6.3) and the repair cost in every damage, which is the annual repair cost under the action of a certain level of wave height (column 9 or $9'$). The sum of column 9 or $9'$ is the expected annual repair cost of a cross-section under the action of all levels of wave heights (equivalent to the average value).

The total investment is calculated as below. If the interest rates are not considered, the total investment is the construction cost of each breakwater cross-section plus the product of the service life of the breakwater and the expected annual maintenance cost. If interest rates are considered, the expected annual maintenance cost should be multiplied by a factor F, which takes the current equivalent values of the future annual payments into account. F is related to the service life n and the annual interest rate i.

$$F = \frac{(1+i)^n - 1}{i(1+i)^n} \tag{6.6.5}$$

If $i=0.08$ and $n=50$ years, then $F=12.23$.

The curve of maintenance cost for different design wave heights in Fig. 6.6.2 is derived on the basis of $F=12.23$. From Fig. 6.6.2, it could be revealed that the design wave height of the breakwater cross-section with the minimum total investment is about 6.75 – 7.0 m, and the corresponding return period is 5 – 26 years. Generally, the design wave heights derived from the optimized design often have the return period of about 10 – 20 years, which is less than the standard of 25 – 50 years return period as specified in the design codes or manuals as usual. The mound breakwaters designed in relative low standards are economically reasonable, but they must be maintained frequently.

The design method may not be suitable for important ports where the breakwater damage will affect the use of the port. However, it can be considered for small and medium-sized ports where the initial construction costs are limited, especially when materials and construction equipment are available for frequent maintenance.

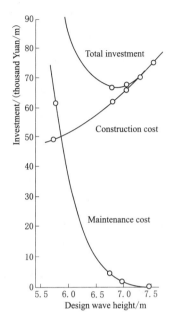

Fig. 6.6.2 Comparison chart of total investment

6.6.2 Optimal design of vertical breakwaters

The difference between vertical breakwaters and mound breakwaters in optimal design lies in that, firstly, the mound breakwater is usually taken as the optimization problem of one variable, while for the vertical breakwater, as the gravity structure, its height and width can both be taken as variables, so it will be more complex for the vertical breakwater. Secondly, for mound breakwaters, when the significant wave height H is greater than the wave height H_{S0}, which corresponds to the no damage criterion ($n\%$ allowable damage rate) of the armour blocks, the embankment will be damaged with different degrees. For vertical breakwaters, when the wave height is greater than the critical wave height of overall instability (sliding or overturning) of the embankment, the overall damage of the embankment will be caused, which means that the damage of the embankment can be considered to be basically the same (of course, different waves will cause the different damage to the foundation bed and the bottom protection in front of the breakwater). Besides, these two types of the breakwater use different design wave heights with different cumulative frequencies, as defined by the relevant design codes, which should also be followed in the optimal design.

The optimized design's method and steps are specified in the following calculation example of the vertical breakwater. The calculated water depth of the proposed breakwater is $d=16$ m. According to the statistical analysis of the field wave observation data, the wave height H_d of different return periods and the corresponding wavelength L are obtained (listed in columns 1, 2 and 3 in Table 6.6.2).

Table 6.6.2　　　　Optimization calculation table of vertical breakwaters width

1	2	3	4	5	6	7	8	9	10
Return period/a	H_d/m	L/m	H_c/m	b_1/m	b_2/m	B/m	Construction costs /(10,000 Yuan/m)	The damage probability P	Total investment /(10,000 Yuan/m)
50	8.7	139	9.4	10.54	8.08	10.6	3.09	2×10^{-2}	4.60
100	8.9	147	9.5	11.19	8.34	11.2	3.21	1×10^{-2}	4.00
500	9.3	162	9.7	12.68	8.93	12.7	3.50	2×10^{-3}	3.67
1,000	9.5	175	9.9	13.48	9.22	13.5	3.65	1×10^{-3}	3.74
5,000	9.9	191	10.0	15.17	9.83	15.2	3.98	2×10^{-4}	4.00

Note: The total height of the breakwater $h=19.0$ m.

Before calculating the wave force acting on the breakwater, whether the wave breaks in front of the breakwater will be firstly determined. According to the results of some flume experiments, the critical original wave height before the vertical breakwaters without wave breaking is

$$H_c = 0.109\tanh\frac{2\pi d}{L} \qquad (6.6.6)$$

Eq. (6.6.6) is applicable to the case of subgrade bed or low bed. The critical wave height values calculated from Eq. (6.6.6) are given in column 4 of Table 6.6.2. As can be seen from Table 6.6.2, each line satisfies $H_c > H_a$, so the wave force can be calculated by standing waves. It should be pointed out here that the above criteria for discriminating the wave breaking are slightly different from those in *Code of Hydrology for Harbour and Waterway* (JTS 145—2015). According to the code, when it is subgrade bed or low bed, $H_c = (0.5 - 0.55)d$. While H_c in column 4 of Table 6.6.2, varies between $(0.59 - 0.63)d$ with d/L according to the formula.

According to the use requirements, the height of the top of the embankment above the calculated tidal level shall be at least 3 m, and when there is the rubble mound foundation of 1.5 m thickness above the seabed, the minimum height of the embankment shall be 17.5 m. The total height of the breakwater $h=19.0$ m is used to optimize the width of the breakwater b.

The stability conditions against sliding along the foundation bed surface of the embankment are

$$K_1 = \frac{(G-P_u)f}{P} \qquad (6.6.7)$$

Where, K_1 is sliding resistance safety factor; P is wave force acting on the embankment body; P_u is wave buoyancy uplift force acting on the embankment bottom; f is the friction coefficient of the embankment body along the foundation bed surface; G is the self-

6.6 Optimal design of breakwaters

weight of the embankment body per unit length, calculated by the following equation.

$$G = (\gamma_1 h_1 + \gamma_2 h_2) b \tag{6.6.8}$$

Where, b is the width of the embankment; h_1 and h_2 are the heights above and below water surface of the embankment body respectively, see Fig. 6.6.3; γ_1 and γ_2 are the average gravity above and below the water surface of the embankment body respectively.

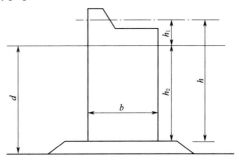

Fig. 6.6.3 Calculation diagram of vertical breakwaters

In the optimal design, $K_1 = 1.0$, because in this way, the corresponding failure probability can be calculated. From this it can be deduced that:

$$b = \frac{\dfrac{P}{f} + P_u}{\gamma_1 h_1 + \gamma_2 h_2} \tag{6.6.9}$$

The stability conditions against overturning of the embankment are

$$K_2 = \frac{(\gamma_1 h_1 + \gamma_2 h_2) b^2}{2 M_0} \tag{6.6.10}$$

Where, K_2 is overturning resistance safety factor; M_0 is overturning moment of P and P_u to point O (Fig. 6.6.3). Similarly, from $K_2 = 1.0$, it can be deduced that:

$$b = \left(\frac{2 M_0}{\gamma_1 h_1 + \gamma_2 h_2} \right)^{1/2} \tag{6.6.11}$$

In order to differentiate, the embankment widths derived from Eq. (6.6.9) and Eq. (6.6.11) are expressed as b_1 and b_2, respectively, in Table 6.6.2 (columns 5 and 6), while the final determined embankment width is expressed as b (column 7).

Once the embankment widths b for different return periods of wave action are determined, the structural design of the embankment body can be carried out. In this example the reinforced concrete caisson scheme is used. The construction cost is then derived from the quantities and unit prices of each item (column 8 in Table 6.6.2).

The following contents describe how to determine the repair cost. When the vertical breakwater is in good integrity, it can be considered either no damage or overall damage under the wave action. After the overall damage of the vertical breakwater, it is necessary to remove the damaged section of the embankment and build the new section. Therefore, the so-called "maintenance cost" will be more expensive than the construction cost. As a rough estimate, it is assumed that each repair cost is twice the construction cost. The expected annual maintenance cost will be equal to $2CP_D$ and P_D is the annual damage probability of the breakwater which is the reciprocal of the return period (column 9 of Table 6.6.2). Because the maintenance cost is paid during the service life of the breakwater after construction, if the interest rate factor is considered, the current value coefficient F should be introduced into the current investment. The formula of F is shown in Eq. (6.6.5), so the total investment is

$$C' = (l + 2P_D F)C \qquad (6.6.12)$$

Assuming the annual interest rate $l = 0.08$, service years of the structures $n = 50$, it could be deduced that $F = 12.23$ from Eq. (6.6.5). The total investments obtained from Eq. (6.6.12) are included in column 10 of Table 6.6.2, which completes the optimization of the total height of the breakwater $h = 19.0$ m. The total investments for different embankment widths b are plotted as the curve, and the optimal embankment width of 12.7 m can be derived in Fig. 6.6.4.

Assuming the embankment heights of 17.5 m, 20 m, 21 m, 22 m, and 23 m, etc., the corresponding optimal embankment widths with the minimum total investment can be obtained as 14.8 m, 11.6 m, 10.7 m, 10.2 m, and 9.8 m respectively. The minimum total investments for different embankment heights h are plotted as the curve, and the optimal embankment height of 21 m can be derived in Fig. 6.6.5. And the $h = 21$ m and $b = 10.7$ m are the final results obtained from the secondary optimization.

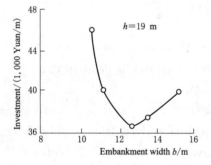 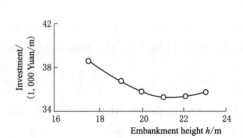

Fig. 6.6.4　Optimization of embankment width　　Fig. 6.6.5　Optimization of embankment height

From Table 6.6.2, it can be found that the wave return period corresponding to the optimum embankment width $b = 12.7$ m at $h = 19.0$ m is 500 years. The wave return period corresponding to the optimal cross-section of $h = 21$ m and $b = 10.7$ m is also 500 years. The wave return period for the vertical breakwater according to the optimal design is larger than the wave return period (e.g., 50 years) determined by the relevant design code. It is different from the result of the optimal design for the mound breakwater. However, since the safety factor is taken as 1.0 in the optimal design of the vertical breakwater, there is not much difference between the optimal cross-section and the results obtained from the usual design criteria ($K_1 = 1.2 - 1.3$) in general.

Exercise

Q1　Fill in the blanks: The layout types of breakwater can be divided into (　　),

( ), ( ).

Q2　Short answer: What's the function of the breakwater?

Q3　Multiple choices: The main structural forms of breakwater are ( ).

A. Mound breakwater　　B. Vertical breakwater　　C. Mixed breakwater

D. Permeable breakwater　　E. Floating breakwater　　F. Hydraulic breakwater

Q4　Short answer: What are the characteristics of the mound breakwater? And what's the application scope of the mound breakwater?

Q5　Short answer: Write down various structural forms of the mound breakwater as the examples.

Q6　Short answer: Briefly introduce the application scope of various structural types of the mound breakwaters.

Q7　Multiple choices: The common concrete block shapes of mound breakwater are ( ).

A. Tetrapod　　　　　　　　B. Four-legged hollow block

C. Twisted I-shaped block　　D. Accropode block

Q8　Short answer: Please briefly introduce the characteristics and application scopes of the vertical breakwater.

Q9　Multiple choices: The structural forms of vertical breakwater are ( ).

A. Gravity vertical breakwater　　B. Pile type vertical breakwater

C. Energy dissipation breakwater　　D. Baffle permeable structure

Q10　Short answer: Please briefly describe the wall body structure, superstructure and foundation bed structure of gravity vertical breakwater.

Q11　Multiple choices: Find out the following errors in describing the types, characteristics and applicable conditions of square vertical breakwater. ( )

A. The types of wall block are ordinary block (normal placed blocks, sloping placed blocks), giant block and wave dissipation block.

B. The advantages are as follows: the wall body is strong and durable, the underwater installation and diving workload is less, the construction progress is fast, and it can withstand large waves.

C. The disadvantages are large self-weight, large foundation stress, large amount of concrete, poor overall performance of embankment body, easy to deform with foundation settlement, and sensitive to uneven settlement.

D. The applicable conditions are as follows: the wave height is not large during the construction period, the on-site crane equipment capacity is large and the foundation is solid.

Q12　Judgment question: The caisson wall of vertical breakwater mainly has rectangular, circular and concrete caisson with energy dissipation chamber. Judge the correctness of the following characteristics:

(1) The embankment body has good integrity, small installation workload on water, and there is no need to fill sand in the box, so the cost is low.

(2) The prefabrication of caissons and underwater construction need corresponding sites and equipment, and there should be a channel with sufficient water depth.

(3) Large crane equipment is required and construction progress is slow.

(4) The box wall is thin, and once it is damaged, it is difficult to repair.

Q13 Short answer: Please introduce the characteristics of rectangular caissons, circular caissons and caissons with energy dissipation chambers.

Q14 Vocabulary explanation: Large diameter cylindrical vertical breakwater.

Q15 Fill in the blanks: The pile type vertical breakwater has ( ) and ( ) forms, and the steel sheet pile lattice structure.

Q16 Short answer: Make a gentle slope on the sea side of the superstructure of the vertical embankment, just like cutting off a corner from the vertical wall, also known as the chamfered vertical breakwater. Briefly introduce its characteristics.

Q17 Short answer: How to reasonably select the section structure of the mound breakwater?

Q18 Fill in the blanks: Except for special requirements, the design wave height of the mound breakwater adopts the wave height with a return period of ( ) or ( ) and a cumulative frequency of ( ) but not exceeding the limit wave height in shallow water.

Q19 Short answer: How to determine the crest elevation of mound breakwater?

Q20 Fill in the blanks: The width of the top of the mound breakwater ( ) the design wave height, and the structure should be able to place at least two rows of artificial blocks in parallel or three artificial blocks of random placement to ensure that the top of the breakwater is not damaged by waves. A large number of investigations and tests show that the slope embankment section with the crest width ( ) the design wave height is unstable.

Q21 Fill in the blanks: The elevation of the top surface of the riprap prism outside the mound breakwater should be avoided in the strong wave action area ( ) above and below the water surface within the design wave height range. The width and thickness of the top surface of the prism shall be determined according to the water depth in front of the embankment and the section size, and its width shall not be less than ( ), and its thickness shall not be less than ( ). The prism width of the deep-water breakwater should not be less than 5 m and the thickness should not be less than 3 m. The weight of the block stone is the ( ) weight of the armour block stone.

Q22 Short answer: What are the mature principles for determining the position and width of the berm of the mound breakwater?

Q23 Short answer: What guiding principles can be used as design reference for the selection of core materials of mound breakwaters?

Q24　Short answer: In order to ensure the structural safety of the head of the mound breakwater, the head of the breakwater is often specially designed. Briefly introduce some principles of the structural design of the head of the breakwater.

Q25　Short answer: What are the type selection principles of vertical breakwater for design reference?

Q26　Fill in the blanks: Unless otherwise specified, the design wave height of the vertical breakwater refers to the wave height with the return period (　　) and the cumulative frequency (　　) of the wave train, but does not exceed the limit wave height in shallow water.

Q27　Short answer: What are the design principles for the crest elevation of vertical breakwater, the crest elevation of wall body (the crest elevation of caisson or the uppermost block), and the crest elevation of foundation bed?

Q28　Fill in the blanks: In addition to the height of the foundation bed of the vertical breakwater meeting (　　), the rubble mound foundation shall be controlled within the width of the dike body of (　　), and the bottom width of the subgrade bed shall not be less than the bottom width of the vertical breakwater wall plus the thickness of the foundation bed of (　　).

Q29　Fill in the blanks: In principle, the width of the vertical breakwater is determined by stability calculation, which must meet the requirements of (　　), (　　), and (　　). In the initial design, it can be taken as $B=0.8\times$ embankment height.

Q30　Short answer: What are the main advantages of floating breakwater compared with traditional bottom seated breakwater?

Q31　Multiple choices: What are the main disadvantages of floating breakwaters? (　　)

A. Requires a variety of technical equipment to meet the dissipation of design waves.

B. Cable and anchorage are easy to be damaged under wave action.

C. If the cable is damaged, the floating body floats freely, endangering the ship, coast and offshore structures.

D. The field maintenance of floating breakwater is much more difficult than that of fixed breakwater.

Q32　Judgment question: The design of the floating breakwater should ensure that the natural frequency of the floating breakwater is larger than the wave frequency.

Q33　Short answer: Please describe the wave dissipation mechanism of floating breakwater briefly.

Q34　Multiple choices: (　　) are not floating breakwater types.

A. Concrete caisson type　　B. Tire type　　C. Vertical type　　D. Foam type

Bibliography

[1] ABBOTT B M, PRICE A W. Coastal, estuarial and harbour engineer's reference book [M]. Boca Raton: Taylor and Francis, CRC Press, 2014.

[2] CHI Z, et al. Coastal hydrodynamics morphodynamics [M]. 2nd. Beijing: China Communications Press, 2022.

[3] COLLINS Thomas J. Ports'01: America's ports: Gateway to the global economy [M]. Reston: American Society of Civil Engineers, 2001.

[4] DOMINIC R, ANDREW C, CHRISTOPHER F. Coastal engineering: processes, theory and design practice [M]. 3ed. Boca Raton: CRC Press, 2018.

[5] EDGE Billy L. Coastal engineering 2000 [M]. Reston: American Society of Civil Engineers, 2001.

[6] EL-HAWARY F. The ocean engineering handbook [M]. Boca Raton: Taylor and Francis, CRC Press, 2000.

[7] FODA A M. Seafloor dynamics and coastal engineering applications [M]. Boca Raton: Taylor and Francis, CRC Press, 2010.

[8] HAIDA M. Concepts of marine engineering [M]. Singapore: Tritech Digital Media, 2018.

[9] HARLEY Mitchell D, et al. Extreme coastal erosion enhanced by anomalous extratropical storm wave direction [J]. Scientific Reports, 2017, 7 (1-4): 6033.

[10] HE K, HUANG T, YE J. Stability analysis of a composite breakwater at Yantai port, China: an application of FSSI-CAS-2D [J]. Ocean Engineering, 2018, 168: 95-107.

[11] WANG Y J, SCHERIBER L G M. Developments in offshore engineering: Wave phenomena and offshore topics [M]. Houston: Gulf Professional Publishing, 1999.

[12] JHA Ramakar, et al. River and coastal engineering: hydraulics, water resources and coastal engineering [M]. Cham: Springer International Publishing, 2022.

[13] CHIEN C H. Mechanics of Coastal Sediment transport [M]. Singapore: World Scientific, 1992.

[14] KAMPHUIS W J. Introduction to coastal engineering and management [M]. 3rd. Singapore: World Scientific Publishing Company, 2020.

[15] KIM Young C. Coastal and ocean engineering practice [M]. Singapore: World Scientific Publishing Company, 2012.

[16] KIM Young C. Design of coastal structures and sea defenses [M]. Singapore: World Scientific Publishing Company, 2014.

[17] KIM Young C. Handbook of coastal and ocean engineering [M]. Singapore: World Scientific Publishing Company, 2009.

[18] KOUTITAS C, SCARLATOS D P. Computational modelling in hydraulic and coastal engineering [M]. Boca Raton: Taylor and Francis, CRC Press, 2015.

[19] LIU Philip L F. Advances in coastal and ocean engineering [M]. Singapore: World Scientific Publishing Company, 2000.

[20] LORENZO M, I M V, JEAN-Francois P, et al. Global long-term observations of coastal erosion and accretion [J]. Scientific Reports, 2018, 8 (1): 12876.

[21] MAHDI S, NAVID M M, REZA M C. Numerical simulation of waves overtopping over impermeable sloping seadikes [J]. Ocean Engineering, 2022, 266: 5.

Bibliography

[22] MARZEDDU A, OLIVEIRA C T, SÁNCHEZ-ARCILLA A, et al. Effect of wave storm representation on damage measurements of breakwaters [J]. Ocean Engineering, 2020, 200: 107082.

[23] MASARU M, SHINJI S. Proceedings of coastal dynamics 2009: impacts of human activities on dynamic coastal processes (with CD-ROM) [M]. Singapore: World Scientific Publishing Company, 2009.

[24] MCKEE J S. Coastal engineering 2008: (In 5 Volumes) [M]. Singapore: World Scientific Publishing Company, 2009.

[25] MEI Chiang C. Applied dynamics of ocean surface waves [M]. Singapore: World Scientific Publishing Company, 1989.

[26] MELBY Jeffrey A. Coastal structures 2003 [M]. Reston: American Society of Civil Engineers, 2004.

[27] MOHAN Ram K, MAGOON Orville, PIRRELLO Mark. Advances in coastal structure design [M]. Reston: American Society of Civil Engineers, 2003.

[28] PHIL D. Modeling coastal and offshore processes [M]. Singapore: World Scientific Publishing Company, 2007.

[29] PING W, D J R, JUN C. Coastal sediments 2015: the proceedings of the coastal sediments 2015 [M]. Singapore: World Scientific Publishing Company, 2015.

[30] PRABU P, BHALLAMUDI M S, CHAUDHURI A, et al. Numerical investigations for mitigation of tsunami wave impact on onshore buildings using sea dikes [J]. Ocean Engineering, 2019, 187: 106159.

[31] REEVE D, CHADWICK A, FLEMING C. Coastal engineering [M]. 3rd. Boca Raton: Taylor and Francis, CRC Press, 2018.

[32] ROBERT M. Sorensen. Basic coastal engineering [M]. Boston: Springer, 2006.

[33] SWAPNA M, KIRAN K G, ABHRA C. Environmental oceanography and coastal dynamics [M]. Cham: Springer International Publishing, 2023.

[34] T W B. Coastal dynamics [M]. Singapore: World Scientific Publishing Company, 2012.

[35] PANG T Z, WANG X Q, TOYIN G A, et al. Coastal erosion and climate change: a review on coastal-change process and modeling [J]. Ambio: A Journal of Environment and Society, 2023, 52 (12): 2034-2052.

[36] TOMOYA S. Coastal processes: concepts in coastal engineering and their applications to multifarious environments [M]. Singapore: World Scientific Publishing Company, 2008.

[37] VALLAM S, ANNAMALAISAMY S S. Coastal engineering: theory and practice [M]. Singapore: World Scientific Publishing Company, 2019.